Philosophies of Hospitality and Tourism

ASPECTS OF TOURISM

Series Editors: **Chris Cooper** *(Leeds Beckett University, UK)*, **C. Michael Hall** *(University of Canterbury, New Zealand)* and **Dallen J. Timothy** *(Arizona State University, USA)*

Aspects of Tourism is an innovative, multifaceted series, which comprises authoritative reference handbooks on global tourism regions, research volumes, texts and monographs. It is designed to provide readers with the latest thinking on tourism worldwide and in so doing will push back the frontiers of tourism knowledge. The series also introduces a new generation of international tourism authors writing on leading edge topics.

The volumes are authoritative, readable and user-friendly, providing accessible sources for further research. Books in the series are commissioned to probe the relationship between tourism and cognate subject areas such as strategy, development, retailing, sport and environmental studies. The publisher and series editors welcome proposals from writers with projects on the above topics.

All books in this series are externally peer-reviewed.

Full details of all the books in this series and of all our other publications can be found on http://www.channelviewpublications.com, or by writing to Channel View Publications, St Nicholas House, 31–34 High Street, Bristol BS1 2AW, UK.

ASPECTS OF TOURISM: 91

Philosophies of Hospitality and Tourism

Giving and Receiving

Prokopis A. Christou

CHANNEL VIEW PUBLICATIONS
Bristol • Blue Ridge Summit

DOI https://doi.org/10.21832/CHRIST7376
Library of Congress Cataloging in Publication Data
A catalog record for this book is available from the Library of Congress.
Names: Christou, Prokopis A., 1978– author.
Title: Philosophies of Hospitality and Tourism: Giving and Receiving/
 Prokopis A. Christou.
Description: Bristol; Blue Ridge Summit: Channel View Publications, [2021] |
 Series: Aspects of Tourism: 91 | Includes bibliographical references and index. |
 Summary: "This book introduces readers to philosophies of hospitality and
 tourism. It provides insights into classic philosophical concepts and explains how
 these can inform the actions of tourism stakeholders, practitioners, hosts and
 tourists. The discussion of philanthropy is a strength of the book and will be
 important in a post-Covid19 tourism industry"— Provided by publisher.
Identifiers: LCCN 2020028431 (print) | LCCN 2020028432 (ebook) |
 ISBN 9781845417369 (paperback) | ISBN 9781845417376 (hardback) |
 ISBN 9781845417383 (pdf) | ISBN 9781845417390 (epub) | ISBN 9781845417406
 (kindle edition)
Subjects: LCSH: Tourism—Philosophy. | Hospitality—Philosophy.
Classification: LCC G156.5.P55 C47 2021 (print) | LCC G156.5.P55 (ebook) |
 DDC 338.4/791001—dc23 LC record available at https://lccn.loc.gov/2020028431
LC ebook record available at https://lccn.loc.gov/2020028432

British Library Cataloguing in Publication Data
A catalogue entry for this book is available from the British Library.

ISBN-13: 978-1-84541-737-6 (hbk)
ISBN-13: 978-1-84541-736-9 (pbk)

Channel View Publications
UK: St Nicholas House, 31–34 High Street, Bristol, BS1 2AW, UK.
USA: NBN, Blue Ridge Summit, PA, USA.

Website: www.channelviewpublications.com
Twitter: Channel_View
Facebook: https://www.facebook.com/channelviewpublications
Blog: www.channelviewpublications.wordpress.com

Copyright © 2021 Prokopis A. Christou.

All rights reserved. No part of this work may be reproduced in any form or by any means without permission in writing from the publisher.

The policy of Multilingual Matters/Channel View Publications is to use papers that are natural, renewable and recyclable products, made from wood grown in sustainable forests. In the manufacturing process of our books, and to further support our policy, preference is given to printers that have FSC and PEFC Chain of Custody certification. The FSC and/or PEFC logos will appear on those books where full certification has been granted to the printer concerned.

Typeset by Nova Techset Private Limited, Bengaluru and Chennai, India.

Contents

Tables, Figures and Photos vii
Acknowledgements ix
Preface xi
Glossary xiii

Introduction: The Nexus of Tourism, Hospitality
and Philosophy 1

Part A: The Philosophy of *Giving* in Hospitality and Tourism

1 Giving Philoxenia: The Concept of Hospitality 15
    Hospitality, Origin and Domains 15
    Philoxenia: Understanding and Practice 20

2 Giving *Agape*: The Concept of Love 30
    Introducing Love 30
    Offering Love: Sex Tourism and Ethical Considerations 31
    Different Forms of Love? 33
    Empathy and Selflessness as Prerequisites of *Agape*? 34
    Agape and its Actions 35

3 Giving Philanthropy: Private and Organisational Philanthropy
in Tourism and Hospitality 44
    Private Philanthropy in the Tourism Context 45
    Organisational Philanthropy in the Tourism Context 48
    Conceptualising Philanthropy in the Tourism and
        Hospitality Context 53

Part B: The Philosophy of *Receiving* in Hospitality and Tourism

4 Receiving Experiences: Tourist Senses and Emotions 59
    The Role of the Senses and Hedonism in Tourism 59
    Emotions in Tourism 65

5 Receiving Happiness: Tourist Well-being and Psychology 77
    The Notion of Well-being 77
    The Notion of Happiness 79
    Well-being and Happiness in the Context of Tourism 85

6 Over-receiving: Unruly Behaviour, Gluttony
  and Overconsumption in Tourism   94
    Unruly and 'Carnivalesque' Tourist Behaviour   94
    The Role of Overconsumption and 'Gluttony' in Tourism   99

Part C: The Philosophy of Giving and Receiving (Tourism) Places

7 Giving and Receiving Places: The Significance and Spirit of
  Places, and Tourism Development   111
    'Topophilia' as Affection for a Place   111
    'Sense of Place' and *Genius Loci*   112
    The Alteration and Obliteration of a Place's *Genius Loci*
      and the Concept of 'Placelessness'   113
    Regaining the *Genius Loci* of Places   118
    A Diagram Illustrating 'Place' within the Context of
      Tourism   121

8 Giving and Receiving Places: Spiritual Tourism
  and Dark Tourism   124
    Spiritual Sites and Spiritual Tourism   124
    Thanatourism and Places of Atrocity   129
    Current Ethical Issues and Concerns Related to Spiritual
      and Atrocity Places   133

Closing Remarks   142

References   144

Index   183

Tables, Figures and Photos

Tables

Table 4.1	Classical theories of emotion	68
Table 4.2	Key themes of tourism research related to emotions	69

Figures

Figure I.1	The philosophy of tourism: Giving and receiving within the act of travel for tourism purposes	8
Figure 1.1	How high-quality standards of service overlap with the qualities of hospitality	19
Figure 1.2	How organisational culture promotes core commercial values to the detriment of philoxenia	25
Figure 1.3	How tourists' unethical and/or egoistical actions influence philoxenia	26
Figure 1.4	How employees' inappropriate and impolite actions influence the offering of philoxenia	27
Figure 2.1	*Agape* and its actions	36
Figure 3.1	Philanthropy in tourism and hospitality	54
Figure 4.1	Intensity of human interactions and spaces influencing emotional states	72
Figure 4.2	Conceptualising the tourism and emotions nexus	74
Figure 5.1	Well-being and perspectives	80
Figure 5.2	Ways of understanding happiness	83
Figure 5.3	The well-being and tourism experiential matrix	89
Figure 5.4	The nexus of tourism and happiness	91
Figure 6.1	The linkage of enclaved tourism areas, liminality and carnivalesque behaviour	97
Figure 6.2	Tourism and overconsumption	106
Figure 7.1	Place in the context of tourism	122

Photos

Photo I.1	Athens, Greece	5
Photo 1.1	Visitors at Alexander Nevsky Cathedral, Tallinn, Estonia	24
Photo 2.1	Sanctuary, Victoria, Australia	39
Photo 2.2	Giraffe Centre, Kenya	42
Photo 3.1	Christmas charity trees at a resort in Orlando, USA	49
Photo 3.2	Bran Castle, Transylvania, Romania	56
Photo 4.1	Venice resort, Macau	60
Photo 4.2	Smurf village, Belgium	61
Photo 4.3	Seaplane Harbour, Estonian Maritime Museum, Estonia	62
Photo 4.4	Sovereign Hill, Ballarat, Australia	64
Photo 4.5	Landscape, Iceland	66
Photo 4.6	Island of Ithaca, Greece	73
Photo 5.1	Prague, Czech Republic	79
Photo 5.2	Mini-Europe theme park, Belgium	81
Photo 6.1	Tropical beach, Thailand	100
Photo 6.2	Food market, Stockholm, Sweden	101
Photo 7.1	Phuket Island, Thailand	116
Photo 8.1	Mount Athos, Greece	128
Photo 8.2	Old Jewish Cemetery, Prague, Czech Republic	130
Photo 8.3	'Nea Moni', UNESCO World Heritage Site, Island of Chios, Greece	133
Photo 8.4	Piazza di San Pietro, Vatican City	135

Acknowledgements

I wish to express my gratitude to everyone who has supported my efforts to write and publish this book. My students and colleagues – many of whom I regard as friends – at Cyprus University of Technology, University of Central Lancashire, Med High and Louis Travel have inspired me to work hard and strive for excellence. Special thanks are owed to Professor Conrad Lashley, who introduced me to the very depths of the hospitality domain, and to the monks at the Holy Monastery of Panagia Machaira for opening their hearts and their library to me. My gratitude is also extended to my lovely wife, Amalia, and my two beautiful girls, Julia and Maria, for giving me the mental strength to accomplish this hard task.

This book is dedicated to every single person and organisation that strives and struggles for the *agathon* (good) for others and for society.

Prokopis A. Christou

Preface

Practising Plato's (*c*.428–*c*.348 BC) notions of *kalon* (fine) and *agathon* (good) could not be more apposite in today's highly materialistic, competitive and self-centred world, or more relevant to the services and tourism sector. As the title indicates, the primary aim of this book is to introduce readers to the philosophy of 'giving' and 'receiving' in tourism and hospitality. The philosophical approach to issues in contemporary tourism is what makes this book thought provoking. The book provides insights into philosophical concepts and explains how these can inform the actions of tourism stakeholders, practitioners, hosts and tourists. The author aspires to speak to the heart of readers through noble notions such as *agape* (a form of love) and philanthropy. As Aristotle (384–322 BC) said: 'Educating the mind without educating the heart is no education at all.'

The author wrote this book in an attempt to address a wide audience, ranging from experts in the area of tourism, to students, practitioners and anyone who might be interested in tourism. The author hopes to cast light on the rather perplexing and multifaceted nature of tourism in order to help practitioners, students of tourism and general tourism bibliophiles truly to become 'friends of (tourism) wisdom' ('philosophy' literally meaning love of wisdom). For this reason, complicated philosophical and other concepts are explained using carefully selected examples from the tourism and hospitality industry. The author merges findings from empirical studies with instances from the tourism sector, covering different types of hotels, restaurants, airlines, museums, national parks and theme parks. In addition, the author uses extensive quotes from newspaper articles and the websites of official tourism organisations, as well as contemporary cases drawn from the international tourism scene.

The author hopes that this book will be of particular interest to the following readers:

(i) **Tourism/hospitality stakeholders, practitioners, managers, service providers and entrepreneurs**. If the aim of any tourism and hospitality stakeholder is to provide an enriched, holistic, sensual, emotional and above all memorable experience for visitors and guests, while taking into consideration any ethical issues that may arise in the process, then this book should be on their reading list. It will enable practitioners to comprehend the importance of love-driven, philanthropic and good actions towards themselves, their employees, their co-workers, their community and their organisation, while promoting their company image and brand.

(ii) **Academics, students, tourism/hospitality curriculum developers and tourism/hospitality trainers.** This book can be used as a guide for the development of any 'introductory' undergraduate or postgraduate tourism and hospitality modules, or for more specialised modules linked to ethics, sustainability and the sociology or psychology of tourism. The whole book, or sections of it, can be used in training programmes that aim to provide insight into how hospitable actions influence the emotional states of guests, how to impact favourably on the experiences of guests while addressing their emotions and senses, and how to promote actions (philanthropic, societal and environmental) that will impact favourably on the community and the organisation.

(iii) **Anyone who has an interest in tourism, is a traveller, has a passion for travel or is considering a course or a career in tourism/hospitality.** Although this book does not provide information about the essentials and basics of tourism or coverage of technical and operational issues, it should be regarded as useful reading for anyone who has an interest in or passion for tourism. They will acquire a fundamental understanding of what is involved in the simple yet meaningful transactional process of 'giving' and 'receiving' within the context of tourism.

Glossary

Agape	A kind of unconditional love that transcends and persists regardless of circumstances.
Agathon	An aggregate of concepts embracing moral and spiritual virtues, particularly goodness, benevolence and kindness.
Cosmos	The universe seen as a well-ordered whole. The term was first used by the classical philosopher Pythagoras (c.570–c.495 BC).
Genius loci	The presiding spirit of a place, its prevailing character and atmosphere. In classical Rome, it was depicted in religious iconography as a cornucopia (horn of plenty) or a snake.
Gluttony	A term derived from the Latin *gluttire*, to swallow or gulp down. It implies overconsumption and overindulgence in food, drink and/or luxury items.
Kalon	That which is fine; the ideal of physical and moral beauty.
Omotenashi	Japanese term that means to look after guests wholeheartedly.
Philanthropy	Literally 'being a friend of the human species'. Based on the principles of agape (see above), philanthropy is the desire to promote the welfare of others, especially those in need of psychological, physical or economic support.
Philokalia	Literally 'being a friend of what is good', *philokalia* is used to refer to a collection of texts written by spiritual masters between the 4th and 15th centuries. These texts are concerned with themes of universal importance, such as the well-being of the inner life of humans and the awareness of (our) passions.
Philosophy	Literally 'being a friend of wisdom', philosophy is the study of the fundamental nature of knowledge, the universe and existence.

Philoxenia	Literally 'being a friend to the stranger', accommodating and comforting a guest based on the principles of *agape* (see above).
Topophilia	Literally 'loving a place'. A strong sense of a specific place.
Virtues	Behaviour that exhibits high moral standards. Particular virtues, such as humility, meekness, charity and love, have been noted to act against certain traits of character that may harm individuals and others, such as greed, pride, deceit and despair.
Weltanschauung	A German term from *Welt* (world) and *Anschauung* (view) which implies a particular philosophy and view of life; the apprehension of the world from a specific standpoint.

Introduction: The Nexus of Tourism, Hospitality and Philosophy

> Nothing endures but change.
> Heraclitus, *c*.535–*c*.475 BC

Philosophy derives from *philosophia*, which means 'being a friend of wisdom (a *philos* of *sophia*)' and, as Socrates once said, 'the only true wisdom is in knowing that you know nothing' (Durant, 2012). Wisdom may be understood as the quality of possessing the triptych of knowledge, experience and good judgement. Philosophy entails the study of fundamental questions about existence, the physical world, knowledge, reason, values and ethics. There are different divisions of philosophical enquiry, such as natural or physical philosophy (from *physis*, nature), and metaphysical philosophy ('after the physics', *meta-physica*), which concerns existence, God and moral philosophy. The importance and deeper meaning of philosophising (including within the context of tourism) has been acknowledged and emphasised by academics:

> [P]hilosophising is the ability to extract ourselves from the busy, engaged world of making and doing things, to disengage and to pause for reflection and thought especially about meaning and purpose. (Tribe, 2009: 5)

There is a very substantial body of knowledge regarding philosophy and philosophical enquiry by philosophers, ancient and modern, often from very different perspectives, regarding the cosmos, life and existence. Presenting even a part of these is beyond the scope of this book; instead, here is the author's own personal selection of quotes by ancient philosophers for consideration and personal reflection. Even after thousands of years, these sayings continue to be of relevance to academics, tourism stakeholders, students and tourists.

(a) **On knowing ourselves and the results of our actions**:

> The content of your character is your choice. Day by day, what you choose, what you think and what you do is who you become. (Heraclitus *c*.535–*c*.475 BC)

Silence is better than unmeaning words. It is better either to be silent or to say things of more value than silence. (Pythagoras *c*.570–*c*.495 BC)

(b) **On constant edification and hard work**:

We are what we do repeatedly. Excellence, then, is not an act, but a habit. (Aristotle 384–322 BC)

Seek to learn constantly while you live. Do not wait in the faith that old age by itself will bring wisdom. (Solon *c*.630–*c*.560 BC)

(c) **On kindness and empathy**:

Do not do to others what angers you if done to you by others. (Socrates *c*.470–399 BC)

(d) **On happiness and gratitude**:

The secret of happiness is not found in seeking more, but in developing the capacity to enjoy less. (Socrates)

He has the most who is most content with the least. (Diogenes *c*.412–*c*.323 BC)

(e) **On self-centredness, greed and modesty**:

Love of power, operating through greed and through personal ambition, is the cause of all these evils. (Thucydides, *c*.460–*c*.400 BC)

To know is to know that you know nothing. That is the meaning of true knowledge. (Socrates)

The classical philosopher Socrates argued that knowledge and wisdom are equated with self-awareness and happiness. In this sense, the truly wise and self-aware person will have knowledge of what is right, will do what is good and will be happy. A person's philosophy of life, moral ideas and behaviour are reflected in their *Weltanschauung*, a German word that means a particular philosophy of life, a particular view of the world. This concept has been applied in various studies in different disciplines (Naugle, 2002; Nielsen, 1993; Weisskopf-Joelson, 1953) and is used to refer to a wide worldview and theory of the cosmos. Christou (2019a) defined the cosmos as the opposite of chaos, in which it acts as an orderly system. The orderly system of the cosmos is dominated by three interrelated dimensions: the physical, the spiritual and the theoretical. The theoretical dimension is reflected in the philosophical cosmology which influences and is influenced by the other two dimensions. It can be argued that a philosophical cosmology entails an ethical and moral dimension of *acting well* towards the tangible and intangible elements of the cosmos.

On the basis that humans have a strong desire to acquire and channel goodness, kindness and virtue towards others and to abolish negative thoughts, passions and acts of evil, philosophers have analysed the role of ethics and morality in life. Ethics often refers to rules that are provided by an organisation or external source (such as an ethical 'code of conduct'), whereas morals are a person's own principles and individual 'compass' regarding right and wrong. Different cultures have different understandings of what constitutes an ethical action (Fennell, 2006); to take one recent example, Tolkach *et al.* (2017) found that ethical decision making differed according to gender, level of education, age and employment status. In any case, history has proven that what is labelled as 'ethical' is not necessarily good. In ancient Greece, as well as in more recent civilisations, it was ethically acceptable to own slaves. In Germany, the Nazis developed a kind of ethnic conscience according to which they had moral obligations only towards their own race, and this helped them to conclude that their humiliation, persecution, deportation and killing of Jews was the 'right thing to do' (Bialas, 2013).

According to Tolkach *et al.* (2017), ethical questions were first addressed by scholars of tourism in the 1990s, with several major works published during the last few decades (including Fennell, 2014, 2015). Authors have provided detailed critical accounts of ethical perspectives in tourism and hospitality (Burns, 2015; Fennell & Malloy, 2007; Höckert, 2015; Innerarity, 2017; Lovelock & Lovelock, 2013). The important role of ethics and morals in the context of tourism and hospitality is now acknowledged, and this is reflected in the extensive attention they have been given by academics, particularly in the following key areas:

- tourism development and sustainable development (Smith & Duffy, 2004; Jovicic, 2014);
- tourism in relation to the environment and animals (Fennell, 2011; Holden, 2003, 2009; Hughes, 2001);
- key players in tourism, such as the supply chain, destination marketers and operators (Campelo *et al.*, 2011; Fennell & Malloy, 1999; Keating, 2009; Sizer, 1999);
- tourism/hospitality research, education, curriculum, pedagogy, teaching and students (Canosa & Graham, 2016; Enghagen, 1990; Hudson & Miller, 2005; Jamal, 2004; Jaszay, 2002; Moscardo, 2010; Ryan, 2005; Yeung, 2004);
- forms of tourism activity, including ethical tourism (Butcher, 2015; Weeden, 2002), slum tourism (Frenzel *et al.*, 2012), dark tourism (Biran & Hyde, 2013), ecotourism (Wight, 1993) and health, medical and 'transplant' tourism (Budiani-Saveri & Delmonico, 2008; Hall, 2011, 2013);
- the perceptions, intentions and behaviour of tourists (Hindley & Font, 2017; Marchoo *et al.*, 2014; Sheppard, 2010; Speed, 2008; Tolkach *et al.*, 2017);
- certain constructs and concepts in relation to ethical tourism, such as 'hedonism' (Malone *et al.*, 2014);
- business ethics and tourism (Holjevac, 2008; Walle, 1995);

- hospitality and ethics/ethical issues (Claviez, 2013; Knani, 2014; Stevens, 2001; Upchurch, 1998; Worth, 2006);
- hospitality and morality/moral issues (Lashley, 2014; Marnburg, 2006; Ogletree, 2003);
- ethical consumption (Weeden & Boluk, 2014);
- morality, morals and tourism (Butcher, 2005; Caton, 2012; Dimitriou, 2017; Grimwood, 2015; Mostafanezhad & Hannam, 2016; Snyder & Crooks, 2010).

The World Tourism Organization established a frame of reference for responsible tourism, the Global Code of Ethics for Tourism (UNWTO, 2019), which sets out the main principles to guide key players. Being responsible in the context of tourism means taking care of human needs and the needs of the millions of animals used in the industry for human enjoyment and benefit (Fennell, 2014). Many scholars (including Tolkach *et al.*, 2017) have stressed that ethical issues in tourism warrant further exploration.

The Philosophy of Tourism: Gaining the Wisdom of 'Giving and Receiving' within the Context of Tourism

Obtaining personal or/and organisational wisdom is not a simple act or process. It requires physical and mental effort, theoretical and practical knowledge and, above all, experience. 'To understand wisdom fully and correctly probably requires more wisdom than any of us have' (Sternberg, 1990: 3). Wisdom has been defined as the capability to deal with critical life experiences in order to facilitate the development of oneself and others (Webster, 2003), and its beneficial outcomes have been acknowledged, at both the personal and organisational level, as in the case of wisdom leadership (Elbaz & Haddoud, 2017; McCann *et al.*, 2014). Gaining wisdom concerning the broad and multifaceted nature of tourism is perhaps easier said than done. Nevertheless, one truth that can be derived from the wisdom and teachings of different philosophers is that the social world may be shaped by the simple yet meaningful transaction of giving and receiving goodness. Many philosophers have discussed the role of good and goodness in human nature and society (Kraut, 1991; McTighe, 1984).

Humans may desire and aspire to receive goodness and may ultimately channel 'good' deeds, actions and words towards others and towards *physis* (nature and the environment) (see Benson, 2013). In a number of his dialogues, Plato affirmed in various ways that what people desire is the good. Similarly, Socrates claimed that no-one wants to go towards what they believe is bad instead of towards the good (Barney, 2010). A person can attain goodness through repeated practice, and virtues can be perceived as 'habits' that make a person good (Kraye, 2007). From ancient times, virtues have mainly been examined from a philosophical and spiritual perspective (Borowitz & Schwartz, 1999; Taushev, 2014), as in Aristotle's time (Hutchinson, 2015). Particular virtues, such as humility, meekness, charity and love, have been noted as counteracting certain character traits that may harm oneself or others, such as greed, pride, deceit and despair (John Climacus, 2019) (see Photo I.1).

The importance of acquiring virtues and channelling them towards others and towards society has been stressed by a number of philosophers, both Eastern and Western, including Aristotle and Confucius (Yu, 1998). Being a friend of good virtue, or 'love of the beautiful, the good' (*philokalia*) is a notion that has been preserved intact throughout the millennia because of its importance at the personal (psychological and spiritual) and societal levels (Bingaman & Nassif, 2012; Palmer & Ware, 2011). In addition to its literal meaning, *philokalia* is used to refer to a collection of texts written by spiritual masters between the 4th and 15th centuries (Nikodemos & Makarios, 1782). These texts are concerned with themes of universal importance: the well-being of the inner life of humans, awareness of the passions, and knowing the power of thoughts to enslave, deceive and blind but also to heal and set free (Cook & Charles, 2010).

From this perspective, the philosophy of giving and receiving in tourism rests within the context of goodness, hospitableness, virtue and morality. Human actions and external events may interfere in the process

Photo I.1 Athens, Greece. Home to the classical philosophers including Plato, Athens has emerged as a destination that offers a variety of tourism experiences ranging from the archaeo-cultural to the gastronomic
Source: Author.

of philoxenia, shaping the dynamic of giving and receiving goodness; in certain cases and under certain circumstances, noble acts may be replaced by acts of anger, misconduct or hostility. People have always travelled for various reasons, including leisure and pilgrimage. More recently, and within the context of tourism, people may travel outside their normal settings in quest of different types of experiences, such as: for medical reasons (Fetscherin & Stephano, 2016); to find adventure (Gardiner & Kwek, 2017); for religious and spiritual purposes (Kilipiris & Dermetzopoulos, 2016); or for gastronomic reasons (Jiménez Beltrán et al., 2016). The tourism industry, with its range of destinations (countries, regions, cities or villages) and stopovers, offers tourists the opportunity to fulfil their personal quests, needs and desires. Among other possibilities, the industry may offer the following facilities:

- means of transport (planes, trains, buses);
- places at which to be hosted (hotels, hostels, peer-to-peer accommodation);
- places to eat and drink (restaurants, bars);
- places and sites for sightseeing, educational purposes, exercise, spirituality, entertainment and amusement (monuments, national parks, heritage centres, monasteries, museums, casinos, theme parks), many but not necessarily all of which are designed or built for the purpose;
- tracks and paths (for walking, trekking or biking).

These facilities are offered along with service provision by employees working in the hospitality industry (including receptionists, waiters and guest relations managers) or in the general tourism industry (for example, at a national park visitor centre or a museum). Tourism stakeholders may purposely target tourists' senses and emotions in their attempt to offer a holistic sensual and emotional experience. The important role of the senses and emotions in people's lives and as part of any human experience has concerned philosophers (Kristjánsson, 2016; Macpherson, 2011) and, more recently, psychologists (Ekman & Cordaro, 2011; Fredrickson, 2001; Lazarus, 1999; Rouby et al., 2016).

The tourism industry may intentionally target the human senses and emotions through the use of different tactics, elements, tools, objects, machines, food, treatments and experiences. Examples include, but are not limited to: luxurious experiences (Chen & Peng, 2014); idiosyncratic sites, such as a museum of the Holocaust (Cohen, 2011); hedonic activities (Io, 2016); unique gastronomic experiences (Kivela & Crotts, 2006); spas (Dimitrovski & Todorović, 2015); utopic experiences (Christou & Farmaki, 2019); souvenirs (Paraskevaidis & Andriotis, 2015; Wilkins, 2011); and the integration of virtual reality technology (Guttentag, 2010; Jung et al., 2016). In terms of human actions, the service offerings of tourism and hospitality providers, hosts and locals may be guided by hospitable principles of kindness, empathy and benevolence (Christou &

Sharpley, 2019; Lashley, 2016), or they may be expressed through anger, hostility and anti-tourism actions or protests (Coldwell, 2017; López Díaz, 2017). As a result, tourists may become recipients of different sensual and emotional experiences (Christou *et al.*, 2018b; Edensor, 2018; Prayag *et al.*, 2017), either positive or negative. These experiences can be shared with others, perhaps on social media platforms (Kim & Fesenmaier, 2017), and carried as memories for a lifetime (Marschall, 2012; Tung & Ritchie, 2011).

For their part, tourists may show appreciation, give thanks to hosts and tourism service providers, pay for the services they were offered or even share their knowledge with locals and offer to help as volunteers (Stainton, 2016). In other cases, tourists may respond in a negative or immoral manner (Moore, 2017), for instance, by exploiting locals and their children (Weiner, 2016) and by using abusive language or even violence towards tourism employees (Chapman & Light, 2017). Regardless of their intentions, tourists produce impacts on destinations, their environment and their inhabitants through the so-called sharing economy exemplified by Airbnb (Hinsliff, 2018), and through mass tourism and overtourism (Colau, 2014; McVeigh, 2009; Milano *et al.*, 2019). Conceivably, a plethora of human and other external actions underpin and shape the notion of tourism as 'giving/receiving'. These may include people's willingness and capacity to channel love (*agape*) towards other people, be they tourists, hosts or employees (Christou, 2018), ethics and ethical considerations (Burns, 2015; Höckert, 2018; Holden, 2018), training-related issues (Dhar, 2015) and technological advancements such as anthropomorphism and service robots (Kuo *et al.*, 2017; Murphy *et al.*, 2017). Figure I.1 explains the philosophy of tourism: giving and receiving within the act of travel for tourism purposes.

Structure of the Book

This book is divided into three main parts:

- The first part examines the 'giving' side (that of tourism and hospitality suppliers, hosts and locals). It includes three chapters that discuss the concepts of hospitality (philoxenia), love and 'philanthropy' within the context of tourism.
- The second part examines the 'receiving' side (that of the tourist). It explores aspects associated with tourist experiences, including their impact on emotions and the senses and on people's psychology and well-being. It also investigates issues linked to the consumption of experiences, unruly behaviour, gluttony and overconsumption.
- The third part discusses the philosophy of giving and receiving 'places' in the tourism context. The first chapter in this part discusses the offering of places for tourism purposes and how the sense of place may

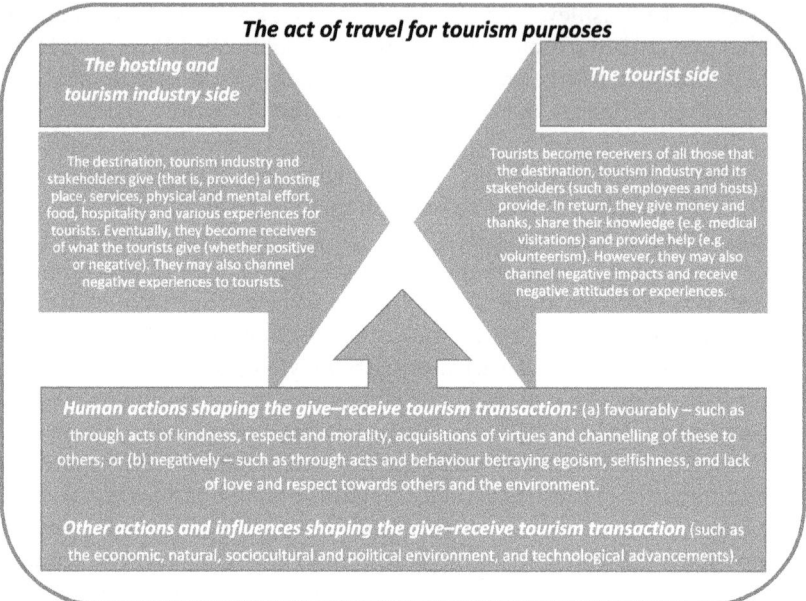

Figure I.1 The philosophy of tourism: Giving and receiving within the act of travel for tourism purposes
Source: Author.

be altered by tourist development and activity. The last chapter examines a range of common cases of moral, ethical and other issues that places (such as those that are important for spiritual and dark tourism) may be called on to address.

Part A. The Philosophy of *Giving* in Hospitality and Tourism

Good actions give strength to ourselves and inspire good actions in others. (Plato, *c*.428–*c*.348 BC)

You cannot be too gentle, too kind … Joy, radiant joy, streams from the face of one who gives and kindles joy in the heart of one who receives. (Seraphim of Sarov, Russia, *c*.1754–1833)

Chapter 1. Giving Philoxenia: The Concept of Hospitality

The origins of hospitality can be traced back through the millennia to the notion of philoxenia, according to which a visitor is perceived and treated as a friend rather than as a customer. Philoxenia is aligned with what can be regarded as the most generous and benevolent form of hospitality: altruistic hospitality. This chapter seeks to explore philoxenia as it is offered now, and to equip individuals, students, academics and practitioners with the knowledge they need to establish some of its core principles.

Chapter 2. Giving **Agape:** *The Concept of Love*

This chapter discusses the broad concept of love in the tourism and hospitality industry, as well as ethical issues associated with love. From a philosophical and sociopsychological perspective, *agape* was introduced to academia in this context by Lee (1977), who viewed it as a form of love coming from an emotionally mature person towards everyone. Through expressions and offerings of love, people may strengthen guest–host bonds, and subsequently guest/host and place affiliations ('topophilia'). This chapter provides insights into the relatively under-researched notion of 'love', specifically in the form of *agape*. It adds to the discourse of a rebirth of social constructive values, such as principles of love, which are being negatively impacted by egoistic and money-oriented attitudes.

Chapter 3. Giving Philanthropy: Private and Organisational Philanthropy in Tourism and Hospitality

Philanthropy was described by Francis Bacon as the habit of doing good and as synonymous with goodness. In recent times, the concept of philanthropy has been a focus of debate within the context of corporate social responsibility (CSR). This increase in importance from an organisational perspective may reflect the apprehension of firms when they realise that their role extends beyond the pursuit of profit. This chapter examines a notion that is important for society but lacking in conceptualisation, aiming to identify the extent to which philanthropy is manifested within the context of tourism (which, unlike many other forms of business activity, is supposedly anthropocentric in character) and to conceptualise philanthropy within the tourism and hospitality context.

Part B. The Philosophy of *Receiving* in Hospitality and Tourism

> No duty is more urgent than that of returning thanks. (Aurelius Ambrosius, *c*.340–397)

> He is richest who is content with the least, for content is the wealth of nature. (Socrates, *c*.469–399 BC)

Chapter 4. Receiving Experiences: Tourist Senses and Emotions

Destinations and tourism/hospitality organisations often attempt to stimulate the senses and emotions of their visitors/guests. As a result, tourists may become receivers of (pleasant/unpleasant) sensual and (positive/negative) emotional experiences. First, this chapter examines the role of human senses and aesthetics within the context of tourist experiences. Secondly, it discusses specific positive and negative emotions such as 'joy' and 'sadness' as part of the tourist experience. The chapter departs from the well-evidenced rhetoric of the role of emotions and their significance in tourism to consolidate research into the link between the two constructs (emotions and tourism). Emerging from this review are the human, spatial and experiential pillars and their interrelation within the tourism and emotion nexus.

Chapter 5. Receiving Happiness: Tourist Well-being and Psychology

It has been suggested that tourists may become recipients of 'joy' and 'happiness', and the last two decades have seen an impressive number of studies dedicated to the relationship of tourism with happiness, positive psychology, well-being and quality of life. Tourist experiences have been found to have a positive effect on a variety of life domains, while the quest for meditative experiences has been seen to contribute to a sense of personal wellness. This chapter discusses the impact of leisure and travel activity on human psychology, considers the experiential matrix of well-being and tourism, and concludes with an analysis of the relationship between happiness and tourism.

Chapter 6. Over-receiving: Unruly Behaviour, Gluttony and Overconsumption in Tourism

First, this chapter discusses issues related to unruly and 'carnivalesque' tourist behaviour at destinations. Secondly, it examines the role of overconsumption and gluttony in tourism, particularly in the hospitality and tourism industry, which has been criticised for the amount of food waste it produces and the impact of its demands on destinations. Overindulgence and overconsumption by tourists may be partly explained by the 'carnivalesque' potential of the destination. Gluttony is one such behaviour that may be linked to tourism overconsumption. From a philosophical perspective, virtues such as prudence, fortitude and abstinence have been advised as ways to counteract gluttony. However, in a consumerist society the promotional tactics of destinations and organisations may trigger overconsumption. This chapter examines whether and the extent to which the tourism and hospitality sector is to be blamed for supporting overconsumption by tourists.

Part C. The Philosophy of Giving and Receiving (Tourism) Places

Chapter 7. Giving and Receiving Places: The Significance and Spirit of Places, and Tourism Development

Destinations are the places in which tourists receive their experiences. Since Tuan (1979) provided the most commonly cited definition of place, tourism researchers have focused on examining the relationship of people with places and what may be referred to as topophilia. Sense of place remains a complex construct. It reflects the ancient and rather indistinct concept of *genius loci*: a location's uniqueness, the soul and spirit of a place. This chapter sheds light on the impact of tourism development and activity on the spirit and *genius loci* of a place.

Chapter 8. Giving and Receiving Places: Spiritual Tourism and Dark Tourism

Places with sacred and historic significance or those that have been the sites of macabre incidents may be offered as tourist destinations. Such places often become a focus of ethical concerns, moral considerations and dilemmas because of their status, importance, meaning or sacredness. Spiritual tourism and thanatourism are two distinct forms of tourism, although they share similarities when it comes to their ethical and moral dimensions. This chapter presents two different case studies within the tourism context: spiritual places and atrocity places. It examines common cases of moral issues, dilemmas and concerns that these places may raise and may be called on to address: the 'anaesthetisation' of visitors, unruly tourist behaviour, and the fact that tourists may take (and share online) 'selfies' on sacred ground or in front of atrocity sites.

Part A

The Philosophy of *Giving* in Hospitality and Tourism

1 Giving Philoxenia: The Concept of Hospitality

> Be kind, for everyone you meet is fighting a hard battle.
> Plato, c.428–c.348 BC

Hospitality, Origin and Domains

Hospitality has existed 'almost since time immemorial' (Brotherton, 2005: 139). The honourable and ethical practice of welcoming and protecting visitors while they are away from home has a tradition almost as long as humanity itself, and is manifested around the world (Blain & Lashley, 2014; Suleri, 2017). Even in times of trouble, the classical Athenians ensured that their city was open to visitors who wanted to benefit from its culture (Thucydides, in Pericles' Epitaph, 430 BC). The act of showing kindness to strangers in the form of benevolent actions towards those in need was a non-optional command of Biblical Scripture (Macris, 2012). The Biblical story of Abraham and Sarah's philoxenia towards three angels that visited them is just one example.

The notion of hospitality continues to be vital for organisations within the tourism sphere (such as airlines, cruise ships, hotels and restaurants), especially those that strive for excellence in service provision. For example, Singapore Airlines won the Skytrax 2019 award for 'Best Airline Cabin Crew', voted for by air travellers. The award recognises the highest all-round performance of cabin staff for: (a) hard service characteristics, such as techniques, efficiency and attention; and (b) soft service characteristics, such as enthusiasm, attitude, friendliness and hospitality. Lashley (2008) proposed a simple and expedient model to understand contemporary hospitality, which has been recognised as a useful tool for investigating the hospitality experience (Christou *et al.*, 2019a; Gehrels, 2017; Ruiter, 2017). Essentially, Lashley's model consists of three domains illustrated in a Venn diagram, with the host–guest relationship and experience as a focal point within the three interrelated domains of the private, the sociocultural and the commercial (discussed below).

The private domain of hospitality

The private domain covers the obligations to act hospitably towards others that are typically learned by individuals (for example, at home). Dekker (2014) argued for the importance of an individual's personality in shaping hospitable attitudes. In the study by Christou and Sharpley (2019), informants who were asked to share their understanding of hospitable/philoxenic people made reference to the 'gift' or the 'charisma' of the person who is hospitable. This gift can be identified through a simple kind gesture and a genuine smile. Additionally, informants expressed the view that people are 'sociable' and learn hospitable actions and behaviour from their home settings or through proper training. The following remarks were made by two informants in the study by Christou and Sharpley (2019):

> ... it is a matter of character to please and even comfort others ... What I mean is that in our culture it is important to be hospitable towards strangers, but you see that some people are not willing to invite you to their homes to have a coffee. ... They are aphiloxenic [*non-philoxenic*].

> It [*philoxenia*] can also be nurtured by our family since our parents might also be naturally philoxenic and they may pass this to their children ... What I mean is that a child learns from his parents to act in the same [*hospitable*] manner towards others ... A child may see his parents taking care of their guests, so he learns to act in the same manner towards his own guests. (Christou & Sharpley, 2019: 44)

The private domain may also overlap with the sociocultural domain:

> You know Northern Irish hospitality ... As soon as someone's in the door the kettle is on. I can't help it. I know it sounds twee, but I almost think of them [*guests*] as friends. (Brownless, 2019)

The sociocultural domain of hospitality

> A country is defined by its people, and the Kenyans are a friendly and proud nation ... Their big smiles, especially on children's faces, and the friendly jambo (greetings) and karibu (welcome) around every corner, goes a long way in making Kenya a welcoming destination! Despite their economic performance, the civil wars and problems that have plagued them, their hearts are always open to outsiders ... a bus carrying school children stopped to help us [*when our car got stuck*] with the teachers getting their hands and clothes dirty, trying to pull our vehicle out! (Sunder, travel and lifestyle writer, 2017)

This domain covers the obligations that different societies place on people to be hospitable. The practice of hospitality is not tied to any particular culture. Cetin and Okumus (2018) discussed local Turkish hospitality, which they grouped under the four themes of 'sociability', 'care', 'helpfulness' and 'generosity'. The gesture of welcoming the stranger is found in

Christian, Jewish and Muslim traditions as a way of expressing love (Reynolds, 2010). *Manaakitanga* can be translated loosely as hospitality; it plays an important role in Māori society in New Zealand, summing up the act of welcoming and sharing. Ducet (in Hughes, 2016), who has been travelling to Afghanistan since the late 1980s, highlighted those areas of the country that are unaffected by the Taliban and where a visitor can enjoy Afghan hospitality, kindness and gentleness.

Furthermore, the concept of *omotenashi* (selfless hospitality) is a cornerstone of Japanese culture, based on the concept that it is a privilege for a host to welcome guests and to make sure that all their needs are met (Tulloch, 2017). In fact, acting on people's needs without being asked to do so is prized by major Japanese organisations; they follow a doctrine of customer service based on *omotenashi* principles, according to which every service comes from the bottom of the heart and is honest and unpretentious. Hospitality is shaped by long-honoured norms as well as by the traditions of a particular culture. Hospitable behaviours range from a welcoming attitude, a caring attitude, helpfulness, generosity and thoughtfulness, to specific hospitable events such as dances, rituals and gatherings (Cetin & Okumus, 2018; Ruiter, 2017; Suleri, 2017).

Religion may also guide hospitable practices. Even in the case of monks and nuns who have retreated to spiritual centres for purposes of peacefulness and above all prayer, the obligation to welcome people and offer hospitality may interrupt their quest for serenity.

> At the monastery of Philotehou, where he [*a monk named Paisios*] had been responsible for the food supplies, he had preferred to do all the work alone, in order to avoid the distraction of interacting with even one single helper. But at 'Stomion,' he found himself among people, since not a single day went by when he did not have visitors. As time passed, more and more people went up to the monastery. He received them with tender love, and was always prepared to offer them some hospitality. Since the bridge of 'Asprolakkos' is visible from the monastery, he could see how many people were crossing it, and by the time they had arrived, he had prepared the coffee, the tea and the 'loukoumi' [*Turkish delight*]. (Holy Hesychasterion, 2018: 178)

The commercial domain of hospitality

> If you are looking for an idyllic break in the countryside with genuine heartfelt hospitality, a farmhouse holiday will fulfil your desires whether in summer or winter. (Salzburgerland, Austria, 2019b)

From medieval inns to contemporary resorts and restaurants, hospitality has long been practised within a commercial domain and, from this perspective, the host–guest transaction is based on what Slattery (2002) characterised as an economic transaction. Destinations (countries and regions) as well as tourism organisations may describe hospitality as 'genuine', 'warm', 'true' and 'heartfelt' in order to stimulate interest. According to Lynch *et al.* (2011), the majority of publications on hospitality emerge from the business sector, which reduces hospitality to an economic activity. Lashley (2008) argued that many industrialised societies do not have a strong obligation to offer hospitality; according to Zarkia (1996), the guest is treated as a client by the host, which means that their relationship is ruled by commercial laws. In commercial settings, hospitality experiences are supplied to paying guests only for as long as they can pay (Lashley & Lynch, 2013). Nevertheless, certain qualities of hospitality (such as a warm welcome and care for the other) may be evident even in commercialised hospitality settings. This is particularly the case for small establishments or venues where, as Lashley (2015a) showed, commercial aspects of hospitality are likely to overlap with domestic ones. This is often (although not necessarily) the case in rural tourist accommodation and restaurant venues.

Research has shown that people working in the tourism and hospitality industry are called on to offer hospitality and to manage their emotions and the display of their emotions. Day (2019) reported that 'princesses' in Disney theme parks have to follow certain rules: they must smile continually, look the part, sound like their character, have exceptionally good posture and point in the proper 'Disney way' when directing a guest. The organisational culture and the managers of many hospitality and tourism enterprises demand a display of positive emotions to impact favourably on the customer experience and, in turn, on the image of the company. This type of display may be effective:

> If what they [*service providers*] express is positive emotions, then they make you feel welcomed and valued as a client. If not, it would be best if they adopt a more neutral approach towards the client but still be polite. The reason is obvious. As a receiver of positive emotions and attitudes, I will also feel and generate similar positive emotions as well. (Tourist, quoted in Christou *et al.*, 2019a: 158)

Employees are required to engage in emotional labour (Beal *et al.*, 2006) when faced with difficult situations such as customer dissatisfaction. In such cases, the labour may involve the display of faked emotions (surface acting); alternatively, employees may work intensively to instil positive feelings in themselves and to channel these feelings to the customers (deep acting). The task becomes extremely challenging when employees have to deal with increased numbers of visitors or unruly behaviour by guests. It can also lead to psychological and physiological fatigue or even exhaustion, when emotional transactions between guests and service providers take place under intense and stressful circumstances (see Simillidou & Christou, 2018).

However, if guests understand that these expressions are 'forced' (for example, by management), they may easily identify the employees' emotional expressions as 'phony' and 'fake'. Guests may perceive service providers as 'over-expressive' if the expressions that are being displayed (whether positive or negative) appear to be exaggerated. Therefore, organisations and their members are urged to shift their focus away from emotional labour and to allow some flexibility, that is, allowing the guest to set the boundaries of the emotionally expressive transaction interface and the employee to follow. This must be done within professional and ethical boundaries. For instance, employees should avoid exaggerated emotional displays (such as excessive enthusiasm) if they perceive that the guest does not value them. Nonetheless, management should intervene to secure the well-being of employees in cases where guests display anger or seek immoral transactions (Christou et al., 2019a).

The adoption of high-quality standards by tourism organisations may overlap with certain qualities of hospitality, thereby helping these organisations offer a hospitable experience (see Figure 1.1). The following examples are from the official websites of an airline based in the United Arab Emirates and a hotel in Zermatt, Switzerland, respectively:

> Your comfort and most importantly your safety are our top priorities at Emirates. We hope you find our world-class service friendly, helpful, attentive and reassuring along your whole journey. ... Our teams have extensive training in safety, security and service to make your flight safe, comfortable and memorable all the way. (Emirates, 2019a)

> But the most important element is you, our guest, with your individual wishes and needs to which we respond with personal service in an informal way. Here you should feel at ease, you should quite simply be yourself. (Omnia, 2019)

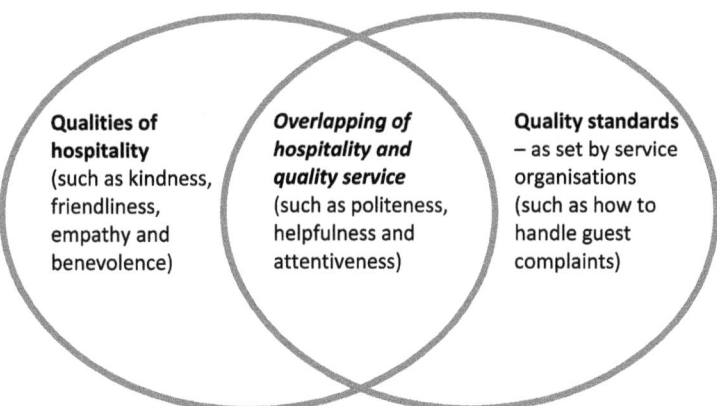

Figure 1.1 How high-quality standards of service overlap with the qualities of hospitality
Source: Author.

Whether the organisation manages to offer a holistic hospitable experience to its guests depends largely on its employees (or hosts) and their willingness to 'go the extra mile' to make guests feel comfortable and truly welcome. In this regard, Lashley (2015b: 1) characterised hospitableness as the 'willingness to be hospitable for its own sake, without any expectation of recompense or reciprocity'. Other researchers (O'Connor, 2005; Tasci & Semrad, 2016) have acknowledged the importance of this sociopsychological phenomenon. For instance, Mody et al. (2019) identified the critical role of hospitableness in facilitating positive customer experience outcomes as well as brand-related outcomes, especially in the context of the hotel experience.

Using an expensive service provider is no guarantee of good service, and a number of hotels and travel organisations choose not to invest in the quality of their services. This allows them to keep their costs low to reduce their rates, thereby increasing demand and sales. A low-cost hotel room or air fare may be tempting for a guest, even if the particular hotel or airline does not offer high-quality service. However, the organisations may be criticised for their corporate culture and attitude towards guests if the basic requirements of service (such as 'politeness') are not met. As the city editor for the *Daily Mail* reported,

> the airlines seem to have given up on any pretence of politeness and customer service, finding even more devious ways to make money from luggage, hand luggage, food and drink. ... Ryanair has, of course, long been loathed by many flyers who hate its extra charges for every last thing, and a corporate attitude that often seems to treat customers as little more than irritating cash cows. (Brummer, 2017)

Philoxenia: Understanding and Practice

Philoxenia essentially means offering friendship (Zarkia, 1996). The term implies acceptance, protection and care for the stranger (Katsaitis & Papaefthimiou, 2019) and that a visitor is perceived and treated as a *philos* (friend). The concept extends beyond the boundaries of commercialised hospitality, since it embraces the active pursuit of ensuring people's comfort, based on principles of *agape* (a form of unconditional love) (Christou, 2018). Although the term phoxenia is Greek in origin, it embraces the conceptual resources to help people from a variety of cultural backgrounds to determine their ethical responsibilities (Nicolacopoulos & Vassilacopoulos, 2004). Philoxenia consists of channelling the positive energy of a person who is in his/her familiar environment to a person who is a stranger in that setting (Katsaitis & Papaefthimiou, 2019). The origins of hospitality can be traced back to Homeric times; a stranger (*xenos*) was treated with sympathy since he/she was away from home, and hosts felt that they ought to provide protection, shelter and food. Acts of philoxenia connected the families of the people concerned, and those links were inherited by their descendants (Homer, 2004).

Ethical hosting obligations
- Offer philoxenia to every passer-by, regardless of their social, financial or political status.
- Treat everyone with the same respect.
- Never raise arms against each other (host against stranger or vice versa), and the same obligation applies to the offspring of both parties.

Material hosting obligations
- Welcome and care for the stranger.
- Offer a meal, a bath and a place to rest and sleep.
- At the end of the stay, goodbyes are accompanied by wishes and gifts.

Later, Paul of Tarsus, known for his teaching on love, urged people to offer philoxenia to others without waiting to be asked (Christou, 2018; Irakleous, 2015; Paul, Apostle, n.d.). Basil the Great (c.329–379), a deeply educated and philanthropic man, spent his wealth on creating a town for those in need. The town (named Vasiliada) included a hospital, an orphanage, a nursing home and a poorhouse. Basil spent most of his time among the poor and offered his help unconditionally; services from all the institutions in his town were provided free of charge to anyone in need. The staff were volunteers who offered their services for the benefit of the community and the society as a whole. (For Basil's approach to life and his teachings, see Athletis, 2018; St Basil, 2016.)

Although there is a spiritual element to philoxenia, it is also aligned with the above-mentioned Japanese concept of *omotenashi* (Tulloch, 2017) and with what can be regarded as the most generous form of hospitality, 'altruistic' hospitality. The latter provides an ideal type of hospitality, since there is no personal (for example, monetary) gain for the host and the offer is an act of benevolence and generosity (Blain & Lashley, 2014; Derrida, 2002; Lashley, 2015b, 2017; Suleri, 2017; Telfer, 2013). For instance, the provision of philoxenia may be considered as a philanthropic act when it involves hosting people who are in need of protection and shelter, without expecting money in return (Christou *et al.*, 2019a).

Guests who have visited an 'Elder' (a spiritually mature person who usually, although not necessarily, lives in a monastic community) typically emphasise the rare, authentic and unconditional nature of the hospitality (philoxenia) they experienced there (Farasiotis, 2005; St Symeon Kolmogkorof, 1998; Lazarus (Moore), 2009; Mantzarides, 2005; Speake & Ware, 2015). In particular, when Elder Paisios withdrew to the countryside, he would leave the door of his hut open to any visitor who wanted to go in and enjoy the dry figs and nuts he had left there (Isaac, 2004).

> ... He [*Elder Gabriel*] counted all adults as children and treated them as if he was their mother. ... If they [*his visitors*] were cold in the hosting area, he sent blankets and bed covers, even his personal winter coats ... Such love is true that only from his own mother someone could receive ... (St Symeon Kolmogkorof, 1998: 198)

According to Christou and Sharpley (2019), deeper meanings of philoxenia involve three elements, all of which must be present:

(a) *The physical element*: for example, the offer of hot tea to a guest on a cold night, without charge.
(b) *The psychological element*: for example, words of kindness and warm-hearted gestures towards guests. Kelly *et al.* (2016) reported that heartfelt acts of hospitality benefited hospital patients, leading to the conclusion that caring about the patient is an important element in the healing process.
(c) *The 'love' element*: for example, leaving our comfort zone in order to pursue actively the comfort of a guest, who is viewed more as a *philos/friend* than a guest.

The ethical and moral dimensions of philoxenia extend to those who are recipients of it, including tourists. Guests are urged to respect those who offer philoxenia, as well as the space (venue and country) in which the philoxenia takes place. For this reason, guests ought to be polite and discreet, respecting the locals and the physical environment and behaving in a decent manner.

> The term 'visitor' brings to mind the discreet knock on the door and the timid: 'Do you allow?' A discreet guest is always a pleasure for the hosts. [S]He considers it an honor to be hosted ... Such a visitor respects the space and things and the boundaries yet does not feel restricted ... This kind of visitor finds it very interesting to follow and albeit even for a short period different rhythms, enriching hence his/her own horizons and opening his/her eyes to new dimensions. It makes sense to him/her not to burden the place which hosts him/her and makes the hosts say 'the trouble of hosting was worth it.' (Liamis, 2019)

Dilemmas, obstacles and challenges of philoxenia

The practice of philoxenia within the largely commercialised world of hospitality is a challenging task which may be accompanied by certain dilemmas:

- To what extent can philoxenia be offered if the organisation is profit oriented and highly commercialised, such as a casino resort hotel? It can be argued that philoxenia cannot be offered or experienced within a profit-oriented environment; nevertheless, philoxenic principles towards others (such as genuine care for guests, employees and colleagues) may well be cultivated within such an organisation. Consider this example from a well-known hotel brand:

> Just as we engage with our guests, we engage with our colleagues to learn about the things they like and are important to them. From their first day of employment, where we learn about their favorite snack and surprise them with it when they return from lunch, to the outpouring of clothes and groceries and even furniture for a colleague whose home burned down. We are constantly on the lookout for ways to surprise and delight! ... Oftentimes in the troubled world we live in, a WOW can come from a genuine smile and a heartfelt, 'How can I help?' delivered to a distressed customer or colleague. Never underestimate the power in WOW! If you create the work environment where co-workers go above and beyond for each other, imagine what they will do for your customers! (Ritz-Carlton, 2015, Response from J. Hargett, Senior Corporate Director, Culture Transformation at the Ritz-Carlton Leadership Center)

- Can philoxenia be quantified and measured? What is a reasonable 'threshold' for an organisation that seeks to adopt a philoxenic approach? One free beverage per guest per day?
- How can an employee offer something (such as a beverage) free of charge to a guest if organisational policies do not allow such actions?
- What if the guest takes advantage of the kindness of the host and 'demands' the offering of philoxenia?
- Will the offering of something free of charge to a guest as a philoxenic gesture raise the expectations of guests in the future? If so, it is highly likely that guests will expect a similar offering on a future visit; if they do not receive it, they will be disappointed.

A number of exogenous and endogenous influences impeding the offering of philoxenia have been identified in the study by Christou and Sharpley (2019): exogenous eco-societal influences, organisational influences, stakeholder influences and guest influences. In addition, employees may influence the offering of philoxenia, and employee influences are included in what follows.

(a) Exogenous influences

Exogenous influences include the following:

- *An economic crisis.* Difficult economic circumstances may deprive service providers (such as hosts) of the opportunity to offer products and services without charge.
- *Overtourism and mass tourism.* Having to deal with large numbers of visitors (see Photo 1.1) may have an adverse impact on philoxenia, resulting in apathy, rudeness or even hostility. In such cases, Doxey's (1975) theory (known as 'Irridex') is highly relevant. Doxey's Irridex

Photo 1.1 Visitors at Alexander Nevsky Cathedral, Tallinn, Estonia. Overtourism makes the offering of philoxenia extremely challenging, even at sacred sites and monasteries which are embedded in traditional norms of welcoming and endurance
Source: Author.

is used to characterise residents' perceptions of tourism in their community and includes 'euphoria', 'apathy', 'annoyance' and 'antagonism'. In the last case (antagonism), tourists greatly outnumber the residents to the point that the locals become antagonistic towards the tourists and engage in rude behaviour. An anti-tourism group in Spain attempted to send a message to visitors in one of the country's most popular destinations (Mallorca) by vandalising tourist rental cars (Bartiromo, 2019). Across Europe, cities such as Venice and Bruges are struggling to cope with overtourism. However, within a relatively short period of time, the outbreak of the COVID-19 virus not only addressed the issue of overtourism but actually caused tourism to dry up completely, even at popular destinations (BBC, 2020; Legorano, 2020). Prior to this, soaring visitor numbers in Amsterdam had prompted the Netherlands tourism board to focus on 'destination management' rather than 'destination promotion' (Fox, 2019).

Hostility towards tourists has reached new heights in recent years. Sightseeing buses in Barcelona have had their tires slashed, cruise ships arriving in Venice are greeted by angry protestors, and anti-tourist graffiti has become commonplace across Europe. 'Tourists go home' appears to be a particularly popular refrain; last year I spotted a slogan on a wall in Lisbon comparing foreign arrivals to a 'zombie invasion.' (Smith, 2019)

- *Automated procedures.* In some service situations, technology and robots have replaced humans. Certain Royal Caribbean cruise ships feature popular 'Bionic Bars' which are operated by robots. The robot bartenders can muddle, stir, shake and strain all types of drinks and cocktail combinations (Royal Caribbean, 2016). In terms of hotel room service, Millward (2018) reported that the days of a hotel member delivering a towel, the morning paper or a meal could become a thing of the past because of the growing use of robots.
- *Increased crime levels and feelings of insecurity.* These may lead to a 'fear' of the unknown and of someone who is a stranger to us. Currently, philoxenia appears to be in sharp conflict with xenophobia and xenophobic tendencies, for example towards migrants (Demetriou, 2012). In fact, philoxenia is the opposite of xenophobia (Hultman *et al.*, 2015); it connotes a relationship far more deep and profound than merely entertaining occasional guests (Kinnamon, 2015).

(b) Organisational and stakeholder influences

Examples include a strong focus on personal gain by owners, a profit-oriented mentality by organisations, and a 'can't be bothered' mentality in general (see Figure 1.2). Lashley (2017) identified a number of motives for offering hospitality, ranging from monetary reasons (commercial hospitality) to more generous motives (altruistic hospitality). Another negative influence is an organisation's choice to implement automated

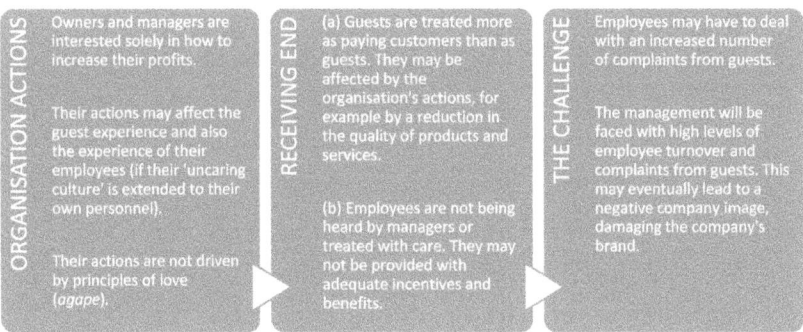

Figure 1.2 How organisational culture promotes core commercial values to the detriment of philoxenia
Source: Author.

technology and procedures in places where human interaction used to be personal, such as check-in machines and robots serving in the hotel bar. With regard to this type of automation, Ritzer (2017) questioned whether what is being offered in many tourism establishments nowadays is actually true hospitality.

(C) Tourist/guest influences

> You must not lose faith in humanity. Humanity is an ocean; if a few drops of the ocean are dirty, the ocean does not become dirty. (Mahatma Gandhi, 1869–1948)

The unethical actions of a minority of tourists need not outweigh the benefit of tourism as a whole. Nevertheless, employees will on occasion have to deal with unruly behaviour (see Figure 1.3). Philoxenia is a two-way process in which two parties (locals/hosts/employees and tourists) must participate; tourists must also consider the host/service providers and treat them with kindness and gratitude. However, shifting values and growing egoistic tendencies on the part of tourists may interfere with the giving/receiving of philoxenia. In such cases, the host or service provider is faced with the challenge (often very difficult) of offering philoxenia.

> Tourists have changed … They are more demanding and wish that things be done in a certain way; their way … They don't like this, they don't like the other thing … It's very hard to please someone who 'demands' your service, your attention and your hospitality … People have become selfish … (Research respondents, quoted in Christou & Sharpley, 2019: 45)

In a report titled '20 things airport employees are too polite to say out loud (but wish they could)', O'Kane (2018) noted that even the most professional airport workers have been pushed close to breaking point by particularly rude travellers but have held back for the sake of keeping the

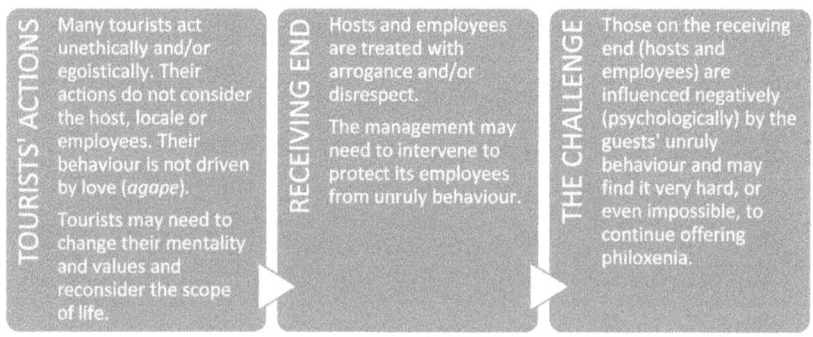

Figure 1.3 How tourists' unethical and/or egoistical actions influence philoxenia
Source: Author.

peace. According to O'Kane, responses that airport employees wanted to make include the following:

- 'Do you realise the pressure we're under?'
- 'Don't blame us for flight cancellations.'
- 'Show some consideration to other passengers.'
- 'You might want to shower before you fly.'
- 'I'm not your waitress/babysitter/servant.'
- 'With that attitude, why should I help you?'

Simillidou and Christou (2018) urged managers to find ways of imposing corporate rules to protect their employees from unethical, unruly, aggressive or hostile behaviour by guests. They argued that management must play an active role in resolving problematic situations. Wang and Groth (2014) noted that employees are better able to cope with aggressive customers and the associated negative consequences if they feel that they have support from management and from their colleagues.

(d) Employee influences

It is debatable whether philoxenia is really being offered to and consumed by guests in cases where employees choose not to align with organisational policies that adopt a pro-philoxenic stance (see Figure 1.4). The following quote is from the manager of an accommodation venue:

> Once I had a receptionist who wouldn't smile to guests and was often impolite to them even in my presence. That's not nice! People come here to have a good time, to relax and the last thing they want to see is a grumpy face, or someone who is rude to them. They will get irritated, angry, or disappointed. You can't have a person like that greeting your guests … Some positions are really important and you must trust them to those who are willing to share the same values with you. (Christou & Sharpley, 2019: 49)

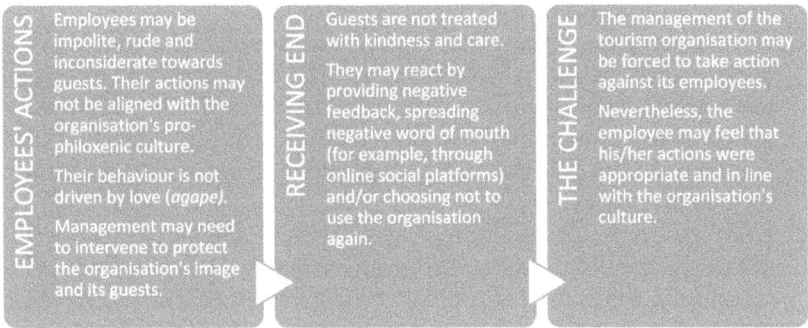

Figure 1.4 How employees' inappropriate and impolite actions influence the offering of philoxenia
Source: Author.

An organisational paradigm shift: From commercial to philoxenic core values

The importance of philoxenia in shaping favourable visitor experiences as well as bringing benefits to organisations is increasingly acknowledged (Katsaitis & Papaefthimiou, 2019). In the commercial domain, as discussed above, it is impossible for profit-oriented organisations to become 'core philoxenic'. Even so, a hospitality and tourism organisation may shift some of its aims in order to adopt a more philoxenic orientation:

> For organizations to become fully philoxenic they must be willing to change ... Change some of their tactics, procedures and the way they view their customers. This means that the owner firstly must be willing to change, the manager and the employees too ... They must not view someone as a euro sign but as a guest that chose to trust to stay with them ... Some people are way too selfish. They don't care about their guests, or even their employees. What they care more is to make more and more money ... to buy a bigger house, to get a better car and go for holidays. (Manager of a rural hotel, quoted in Christou & Sharpley, 2019: 49)

The shift in values from a 'commercial core value' approach to a 'philoxenic core value' approach was illustrated clearly in Christou and Sharpley's study (2019: 40).

(a) *Commercial core values*: profiteering attitudes by service providers/ hoteliers; cultivation of egocentric organisational and personal tendencies; organisational and managerial interests underpinning every action towards the customer; guests treated solely as customers and guests viewing this kind of hospitality as 'less authentic' compared to philoxenia offered with no intention of making a profit.
(b) *People-focused values*: practices that do not betray a profit-oriented culture on the part of the organisation; cultivation of a 'we' rather than an 'I' mentality; guest orientation practices such as 'politeness' promoted and encouraged by the organisation.
(c) *Philoxenic values*: cultivation by the organisation and its members of actions demonstrating love and care for the guest (for example, through kind words and actions); a similar approach adopted by management towards employees; active pursuit (by management and employees) of people's comfort, even where this means leaving their own comfort zone.
(d) *Philoxenic core values*: offering physical and intangible elements without necessarily expecting something in return (such as monetary compensation); application of commercial organisational rules to promote 'philoxenic core values', as in monasteries on Mount Athos (Ayion Oros) where the provision of philoxenia without charge (even for months) is one of the principle *raisons d'être* (Christou, 2016; Speake, 2005).

In this case of philoxenic core values, a pro-philoxenic attitude is practised in which the guest is regarded as a *philos* (friend). Philoxenia is perceived as 'more authentic' by guests than hospitality guided by profiteering practices, since it is less impacted by commercial rules. By offering a gift (even one that is small in value), 'basic' food, or other simple and non-extravagant offerings without charging them to the guest's account, the organisation and its members show their philoxenic core values, which can to some extent be compared to the philoxenia offered in non-profit settings:

> Once I guided two German pilgrims in St Basil's desert. We went to an impoverished kalyve [*hut*]. There the elder received us with kindness. Shortly afterwards he brought to us two tin containers on a wooden tray, filled with rainwater and three figs. He gave us this treat and then offered us dinner from all that he had, some dry bread and olives. The Germans ate the desert's food and were ecstatic, seeing on these desolate rocks between heaven and earth the utmost poverty, possessionlessness and deprivation. (Kotsonis, translated by Mayson & Zion, 1997: 441–442)

2 Giving *Agape*: The Concept of Love

> A heart that is full of love thinks not of itself, but of others ...
> One must be kind, meek, and merciful in one's relationships
> with people.
> Thaddeus of Vitovnica, Serbia, 1914–2003

Introducing Love

> Love is patient; love is kind. It is not jealous. It is not rude. It does not seek its own interests. It is not quick-tempered. (Paul of Tarsus, *c*.15–*c*.68 CE)

Researchers have gathered evidence for the existence of six basic emotions (Ekman *et al.*, 1972), and 'love' is not included on their list. Fromm (2008) regarded love as an interpersonal and creative ability that we must develop as part of our personality, and he did not classify it as an emotion. However, Lazarus (1991) categorised love as an emotion, and Shaver *et al.* (2005) emphasised that love fits the criteria for being a basic emotion. In a study of tourists, Christou (2018) found 'love' to be the emotion most referred to, with almost half of the respondents naming 'love' when asked to give examples of emotions.

Nevertheless, the word 'love' is subject to differing interpretations, and the use and meaning of the locution 'I love you' varies across cultures (Wilkins & Gareis, 2006). People may express love not only for other people but also for a particular job (Kelloway *et al.*, 2010), a specific food item such as chocolate (Poelmns & Rousseau, 2016) or an organisation such as an airline (Thomas, 2015). Others may channel love towards an item or a possession. From a spiritual perspective, Thaddeus (2015) noted the following:

> Some people fall in love with gold and cannot afford their separation from it, or ... their possessions, thus become enslaved ... if someone took these from them, they would be disappointed ... Divine love is infinite and encompasses everything, yet we anchor our love on people and inanimate objects in this world. Our heart is imprisoned by the things of this world, and if they take these away from us, our heart becomes depressed and suffers ... (Thaddeus, 2015: 161)

The importance of 'love' in the context of tourism and hospitality

The importance of love within the tourism and hospitality context has been acknowledged and highlighted at different levels:

- At the personal (tourist) experiential level and at the intrapersonal (host–tourist) level (Christou, 2018; Singh, 2002).
- At a spiritual level: 'Love is the strongest defense available. There are no weapons and power that can be fought with love. In the face of love everything is defeated' (Thaddeus, 2015: 157).
- At a corporate level, as in the case of brand love (Drennan *et al.*, 2015; Wang, Y.C. *et al.*, 2019a); for instance, an empirical study found significant positive effects of (hotel) employee brand love on forgiveness behaviour, supportive voice behaviour and helping behaviour (Wang, Y.C. *et al.*, 2019b).
- At a destination level, as in the case of place brand love (Swanson, 2015). The strong affinity of people for places, also referred to as topophilia (Tuan, 1974), is the result of the subjective and emotional attachment (Creswell, 2004) people have to place. Tourists have been found to embrace emotional relationships with particular settings (Sharpley & Jepson, 2011), and destinations sometimes use the word 'love' in their promotional campaigns. For example, the 'I Love New York' slogan and campaign has been perhaps the most famous tourism marketing campaign in the history of the travel industry (Godfrey, 1984).

Offering Love: Sex Tourism and Ethical Considerations

> Now it is quite true to say that curiosity, exactly like its analogue, lust, never ends and is never satisfied; but man was made for something more than this. He was made to rise, above curiosity and lust, to love, and through love to the attainment of truth. (Seraphim Rose, 1934–1982)

The majority of the tourism and hospitality literature on love is located within a romantic, sex and/or lust perspective, such as travellers in pursuit of romantic affairs with locals (Pruitt & LaFont, 1995), 'love motels' (Alexander *et al.*, 2010) and sex tourism (Cabezas, 2004; Omondi & Ryan, 2017; Ryan & Trauer, 2005). Spencer and Bean (2017) found that hotel workers in Jamaica (both men and women) tend to view male tourists as searching for sex and female tourists as searching for romance. Millions of tourists from India and China visit North American, European and Australian destinations, influenced by media images of Western women as promiscuous and immoral (Bandyopadhyay, 2013). Women on the Kenyan coast, motivated by the desire for a better life, offer 'romantic' experiences to tourists; unfortunately, in the end the women run significant risks such as becoming addicted to the 'easy life' in a way that compromises their ability to earn a living in other ways (Omondi & Ryan, 2017). De Wallen district in Amsterdam is popular among tourists for its sex shops and window brothels, to the extent that the increasing footfall of visitors along with camera phones and social media have turned the district into a spectacle which can be humiliating for the workers there, as the mayor's office stated:

> For many visitors, the sex workers have become no more than an attraction to look at. In some cases this is accompanied by disruptive behavior and a disrespectful attitude to the sex workers in the windows. (Woodyatt, 2019)

Yeoman and Mars (2012) raised thought-provoking points in reference to the use of android prostitutes to address future sex tourism demand, as in Amsterdam's red-light district. On another note, Brazil's far-right President was accused of inciting hatred towards LGBT people for declaring that the country should not become a gay tourism paradise:

> If you want to come here and have sex with a woman, go for your life. But we can't let this place become known as a gay tourism paradise. Brazil can't be a country of the gay world, of gay tourism. We have families. (Reported in Phillips & Kaiser, 2019)

The sexual exploitation of children in tourism

Ethical considerations and policy and legislation options are raised worldwide by the issue of the sexual exploitation of children in tourism, especially in developing nations (Newman *et al.*, 2011). Unfortunately, children in tourist destinations are preyed upon by visitors for sex, with only a few sex tourists being punished (Moloney, 2017). The UN World Tourism Network on Child Protection aims to prevent all forms of child and youth exploitation in the tourism sector (including sexual exploitation and child trafficking). The network provides information about what individuals and organisations can do to prevent child exploitation in tourism. In particular, they advise governments and national tourism authorities as follows:

> Adopt specific legal and administrative measures to address child exploitation in tourism at the local, regional and international levels, including legislation that criminalizes sex with minors and extraterritorial laws that penalise acts committed outside the country of nationality and/or residence. (UN World Tourism Network on Child Protection, 2019)

Bestiality, animal sex tourism and human–animal affection

Bestiality (sometimes referred to as zoophilia) can be understood as sexual intercourse between a person and an animal. In 2015 Denmark banned bestiality, but until then sex with animals was legal provided the animal was not harmed; activists claimed that the previous law was difficult to enforce and was making the country a hotspot for 'animal sex tourists' (BBC, 2015). In 2016 two complainants stating that they were sexually attracted to animals failed in their attempt to get Germany's constitutional court to consider their appeal against laws banning sex with animals (BBC, 2016b).

According to Griffiths' (2016) report on the psychology of bestiality, many zoophiles believe that in years to come their sexual preference will be seen as no different from being gay or straight. Despite criticism and ethical issues arising in connection with bestiality, affection shown to animals and the desire to interact with them for therapeutic purposes within the context of tourism is growing. Mala (2019) reported that cow cuddling (also known by its Dutch name, *koe knuffelen*) is a recent tourism trend, especially in parts of Europe and the United States. The concept is based on animal therapy and invites interaction with farm animals via brushing, petting or even heartfelt chats:

> In today's world, where a lot of communication seems to be through screens, I think people are missing a real connection to being outside in nature … The cuddles and the experience in nature are necessary to stay healthy, mentally and physically … just feeling that from them [*heart rate of cows*] makes you react to that and makes you slow down too. And because you're doing that in nature, it's really quiet … We're not a petting zoo … For us, it's very important that the animals have choices and it's as much their choice to connect with humans as it is for us … usually, the cows love the interactions. And it's not surprising, because cows and horses are social animals. (Vullers, quoted in Maxouris, 2019)

Different Forms of Love?

Scientists recognise the existence of different types of love. For instance, in 'passionate (or obsessive) love', union with the other is associated with fulfilment and ecstasy, whereas separation is associated with emptiness, anxiety and/or despair. 'Companionate love' involves a far less intense emotion, with feelings of deep attachment, commitment and intimacy. Love relationships may involve both types of love (Hatfield & Rapson, 1993a, 1993b; Jungskik & Hatfield, 2004). In an analysis of love, Lee (1977) focused on six distinctly identifiable types of lovers:

(a) the *erotic* lover, who is not anxiously looking for love but is ready for it;
(b) the *ludic* lover, who does not fall in love, is not ready to commit themselves, and feels jealousy;
(c) the *storgic* lover, who expects that love will be an extension of deep friendship towards eventual sexual intimacy and commitment;

(d) the *manic* lover, who shows the same intensity as the erotic lover and manipulates the relationship like the ludic lover, but is usually anxious to fall in love when they are lonely;
(e) the *pragmatic* lover, who combines the manipulation of the ludic lover with the companionship of the storgic lover;
(f) the *agapic* lover, who is more emotionally mature and feels that love is a duty and everyone is worthy of love, despite appearance and merit. 'It [*agape*] has never been better defined than in the words of St Paul to the Corinthians' [referring to the first Epistle] (Lee 1977: 180):

> Love [*agape*] is patient, love [*agape*] is kind. It is not jealous, is not pompous, it is not inflated, it is not rude, it does not seek its own interests, it is not quick-tempered, it does not brood over injury, it does not rejoice over wrongdoing but rejoices with the truth. It bears all things, believes in all things, hopes all things, endures all things.

From a spiritual perspective, there seems to be only one form of love, the kind that Lee (1977) regards as *agape* and that is expressed through heartfelt actions of kindness, benevolence and generosity towards everyone. To camouflage his *agape*-driven intention to over-treat guests, a particular and exceptional host added more sugar or cream to the cups of his guests while exclaiming 'Ooops!' (St Symeon Kolmogkorof, 1998: 198). This kind of love addresses everyone and not the specific other, it does not know how to get angry or exacerbate, and *pathos* (passion) is absent (Isaak, 2011). Of course, this does not imply that *agape* is never experienced in the context of other types of love.

Empathy and Selflessness as Prerequisites of *Agape*?

Empathy is promoted as an emotional prerequisite for cross-cultural understanding. It connects with tourism in a variety of ways (Tucker, 2016) and has been found to reduce antisocial, vengeful, discriminatory and unethical behaviour in service settings (Bove, 2019). The importance of empathy in the tourism context has been reinforced by a number of studies in the field (Costa *et al.*, 2004; Madera *et al.*, 2011; Pera *et al.*, 2019). A study by Umasuthan *et al.* (2017) collected data from 330 hotel guests who had complained about service failures. The results revealed that empathetic dimensions strongly influenced the service experiences of these guests. In the view of Pizam (2015), empathy is necessary for high-quality customer service (although it should be stressed that the author acknowledged that this supposition lacks solid empirical evidence):

> [I]t is my firm belief that while empathy is an important skill that ought to be developed and practiced in all aspects of human life, including interpersonal relations in customer service environments, it is not a skill that is absolutely essential to the delivery of high-quality customer services. (Pizam, 2015: 150)

The link between empathy and love and their psychotherapeutic dynamics towards individuals has been a focus of academic interest for decades (Chessick, 1965; Hart, 2000). Empathy may be considered an expression of love and genuine interest towards another person and his/her burdens:

> People come and tell me their sufferings and my mouth fills with bitterness, as if I ate onions. And when someone tells me that their problem got better or it is solved, I say 'thank God' they gave me a bit of halva [*sweetness*]. When a person is really suffering, I can even die to help [her]him. (Paisios the monk, quoted in Isaac, 2004: 524)

Agape and its Actions

> Every time you smile at someone, it is an action of love, a gift to that person, a beautiful thing. (Mother Teresa, missionary, 1910–1997)

Expressions of *agape* can be fostered by people regardless of their age, beliefs and political/social or economic status. One person who devoted her life to the service of others is (Mother) Teresa, who was recognised by the Indian government in 1962 with the Padma Shri Award for distinguished contribution to society. Her actions and work inspired people and stimulated the growth of 'voluntourism' and work with those in need, including with orphans and leprosy victims (Bandyopadhyay, 2018). Another example is Princess Elisabeth (Feodorovna), the granddaughter of Queen Victoria, who was a German Hessian and Rhenish princess. In 1909 she sold her expensive and impressive collection of jewels and luxurious possessions; with the proceeds she opened the Martha and Mary Convent in the name of 'Mercy and Love'. In the grounds of the convent she opened a hospital, a pharmacy, a school, a library and an orphanage. She worked tirelessly among the poor and the sick and she often visited Moscow's worst slums in order to help alleviate suffering there. Eventually, she was arrested and thrown into an abandoned mine with other prisoners to die (Feodorovna, 2005; Mesa Potamos, 2019).

Figure 2.1 sets out the key features and actions (that is, results) of *agape*, which include acts of philanthropy and philoxenia. The figure also provides paradigms for actions that show a lack of *agape* at the personal, managerial and organisational levels, such as selfishness, greed and lack of respect towards other people (managers, employees, locals, hosts, guests or travellers). Thoughtlessness and lack of respect towards others is clearly illustrated in an incident in which a passenger on a plane was caught using his bare toes to swipe through the in-flight entertainment system, much to the horror and disgust of other travellers (Godfrey, 2019a). In another incident, a beach popular with tourists in the Philippines was forced to close after a woman was spotted burying a nappy in the sand while her mother washed the child in the sea (Godfrey, 2019b).

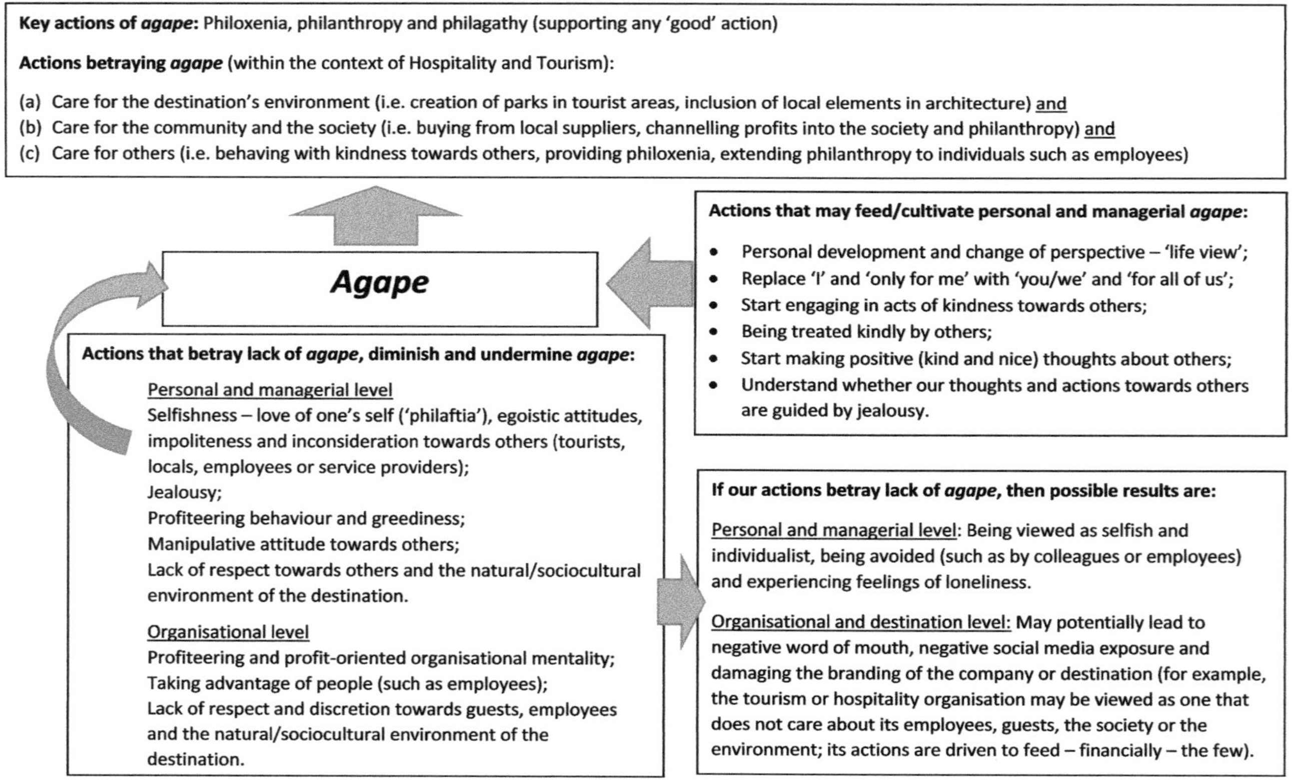

Key actions of *agape*: Philoxenia, philanthropy and philagathy (supporting any 'good' action)

Actions betraying *agape* (within the context of Hospitality and Tourism):

(a) Care for the destination's environment (i.e. creation of parks in tourist areas, inclusion of local elements in architecture) and
(b) Care for the community and the society (i.e. buying from local suppliers, channelling profits into the society and philanthropy) and
(c) Care for others (i.e. behaving with kindness towards others, providing philoxenia, extending philanthropy to individuals such as employees)

Actions that betray lack of *agape*, diminish and undermine *agape*:

Personal and managerial level
- Selfishness – love of one's self ('philaftia'), egoistic attitudes, impoliteness and inconsideration towards others (tourists, locals, employees or service providers);
- Jealousy;
- Profiteering behaviour and greediness;
- Manipulative attitude towards others;
- Lack of respect towards others and the natural/sociocultural environment of the destination.

Organisational level
- Profiteering and profit-oriented organisational mentality;
- Taking advantage of people (such as employees);
- Lack of respect and discretion towards guests, employees and the natural/sociocultural environment of the destination.

Agape

Actions that may feed/cultivate personal and managerial *agape*:
- Personal development and change of perspective – 'life view';
- Replace 'I' and 'only for me' with 'you/we' and 'for all of us';
- Start engaging in acts of kindness towards others;
- Being treated kindly by others;
- Start making positive (kind and nice) thoughts about others;
- Understand whether our thoughts and actions towards others are guided by jealousy.

If our actions betray lack of *agape*, then possible results are:

Personal and managerial level: Being viewed as selfish and individualist, being avoided (such as by colleagues or employees) and experiencing feelings of loneliness.

Organisational and destination level: May potentially lead to negative word of mouth, negative social media exposure and damaging the branding of the company or destination (for example, the tourism or hospitality organisation may be viewed as one that does not care about its employees, guests, the society or the environment; its actions are driven to feed – financially – the few).

Figure 2.1 *Agape* and its actions
Source: Author.

The figure presents exemplars of actions that may be used by people to cultivate their *agape*, such as starting to think positively about others and treat them with kindness.

Figure 2.1 also sets out possible negative results for individuals and organisations if their actions show a lack of *agape*. Examples include the negative social media exposure of an organisation by tourists when they perceive that they have been taken advantage of or deceived. In 2018, a visitor to a café in St Mark's Square in Venice was left feeling stunned after receiving a bill for €43 for two coffees and two bottles of water. The visitor posted a photo of the bill on social media to express his outrage, and the post was shared around 10,000 times. At the same café in 2013, four customers were charged €95 for four espresso coffees with liqueur, and they likewise took to social media to share their indignation (Whitehead, 2018).

Most notably, *agape* tends to initiate actions that show care for others (such as guests), the community/society and the environment. For instance, *agape* for another human being may lead to politeness, words of comfort, kindness and selflessness. Gratitude shown by a guest towards the host or service provider can also be regarded as an expression of *agape*. Gratitude can take many forms, such as a smile showing contentment or a simple, honest 'thank you'. As a form of gratitude, tipping may be appreciated, accepted or rejected by the service provider, depending on the destination and the organisational culture, and it is debatable whether a forced action (such as imposed tipping) should be classified as a genuine expression of gratitude. Tipping is regarded as a basic problem in many service situations (Shamir, 1984), and Ferguson *et al*. (2017) highlighted that knowing where tipping is the norm and where it may cause offence can enhance the service satisfaction experience. Kirchgaessner (2015) noted that in restaurants in Italy a tip is not expected; therefore, it is seen as generous. In Japan and Korea, if someone leaves coins after settling the bill, the waiter may pursue him/her to hand back the cash (Calder, 2015). Tipping is a requirement when eating out in the United States, with an appropriate tip being about 18% of the amount on the bill. However, beliefs about acceptable tipping ranges vary, with some diners feeling that this symbol of gratitude should be earned rather than expected (Ritschel, 2018). Morris (2018a) argues whether America's tipping culture is out of control by questioning why visitors should be forced to reward bad service.

Agape may prevent someone from acting in an immoral manner towards a guest, an employee, a host or even a destination as a whole and its people. In an incident in Bali, two tourists filmed themselves splashing each other with holy water at a temple. The man is shown lifting up his girlfriend's skirt and splashing water on her bottom. Because the incident took place at a temple, it caused particular outrage among the Balinese. In response, the island's governor said that in future any tourists behaving like that should be sent home (Koster, 2019).

Examples of 'lack of love' and lack of care for the other were found in the wake of the sad collapse of the Thomas Cook travel group. Police in Northern Ireland issued a warning that scammers were targeting Thomas Cook customers, pretending to be from their bank, offering to refund deposits and asking them to provide personal details (Belfast Telegraph, 2019). At the other end of the spectrum, some dedicated Thomas Cook staff, finding themselves unemployed, worked for free to help customers and passengers with their travel arrangements (Godfrey, 2019c). A group of stranded passengers raised money to pay the crew who volunteered to work for free on their repatriation flight from Cyprus to the UK (Parikiaki, 2019).

According to Papadopoulos (2014), the main expression of *agape* is philanthropy, which can take the form of providing physical and psychological support to individuals or organisations, such as giving money or things, providing consolation to lonely people, or comforting those who are grieving or experiencing life difficulties. Hence, key actions of *agape* may be philanthropy, philoxenia and, more generally, philagathy, which can be understood as liking and supporting any good actions and deeds towards others (that is, what classical philosopher Plato referred to and supported as *kalon*, that which is fine). Plato appropriated this notion, along with the notions of the good and the just, as a key object for human understanding and as one of the properties of the universe and existence. Lack of *agape* may lead to carelessness, impoliteness and unethical behaviour; egoistic love of one's self may lead to selfish actions:

> When we do not exclude ourselves from our love, our love as great as it is, it is not pure ... it is rancid. But when we take ourselves out of it, then our love is bright. When in our love there is our self, it means that selfishness is in love. But selfishness and love do not go together. Love and humility are two twin brothers tight-lipped. [S]He who has love also has humility, and [s]he who has humility has love. (Paisios the monk, quoted in Isaac, 2004: 526)

Thus, personal and managerial actions that may possibly betray lack of *agape* are selfishness, inconsideration and a manipulative attitude towards others. Taking advantage of employees and possessing a core profit-oriented mentality are behaviours that may show a lack of *agape* at the organisational level. Likewise, 'surveillance' raises ethical concerns regarding human rights and freedom; if used inappropriately by employers, it may be considered as a lack of *agape*. Saner (2018) reported that new technology including wristband trackers, sensors and even microchip plants enables employers to watch staff in more and more intrusive ways, monitoring computer screens, toilet breaks and even emotions.

The concept of *agape* in the context of tourism was explored by Christou (2018), who identified deeper meanings of *agape* as well as certain actions that may underpin or undermine it. Christou's study distinguished three main layers at the heart of the concept: starting from the first layer, we move towards the core of the notion, which consists of anthropocentricity and deep love for the 'other'.

(a) The first layer is caring for the destination and its environment (see Photo 2.1). Locals, tourists and organisations are called on to respect and protect the destination's cultural and natural environment and not damage it. The 'Lake Wanaka Tree' is a natural landmark of New Zealand, but it has been damaged by overenthusiastic visitors who climb the tree, straining the branches and causing them to snap (Hallinan, 2018b). In Kenya in 2016, in an attempt to shock the world into protecting endangered elephants, more than 100 tons of ivory were burned, including tusks, ivory sculptures and rhino horn confiscated by the Kenyan authorities. The ceremony was designed to highlight the decline in Africa's wild elephant population and the dramatic impact of poaching (Smith, 2016). Similarly, Lucas (2019) reported that wildlife authorities were warning tourists against feeding native animals at one of Western Australia's most famous beaches, in effect 'killing them with kindness' in search of the perfect selfie. Many tourists were using human food to lure kangaroos for the perfect holiday snap; as a consequence, the kangaroos became ill from eating unsuitable food, and once they were used to people feeding them they could become aggressive if not fed. The animal protection organisation Animals Australia (2019) urged tourists to travel with kindness and

Photo 2.1 Sanctuary, Victoria, Australia. Sanctuaries, such as this one in Victoria, Australia, may receive less criticism than zoos, since they typically take care of animals that have been abused, neglected or abandoned
Source: Author.

questioned whether tourists spending their holiday money were unwittingly supporting animal cruelty. The organisation aimed to raise awareness of the hidden cruelty behind elephant rides, of zoos that do not meet even the most basic needs of animals with some simply 'displayed' in barren cages, and of certain local cuisines that are based on horrific cruelty, such as *foie gras* and live octopus on a plate.

(b) The second layer is caring for society and its members. This implies organisations genuinely acting for the sake of the community, the society and its members, including tourists. For example, in an effort to address the deadly human traffic jams on Mount Everest, the Nepalese authorities proposed new safety rules to reduce the number of permits issued for climbing the world's highest peak; tourism companies would be required to have at least three years' experience of organising high-altitude expeditions before being allowed to lead climbers on Everest (Sharma & Schultz, 2019). More generally, when organisations undertake corporate social responsibility (CSR) initiatives, this should be done in a way that demonstrates genuine interest in society rather than being driven by marketing objectives. For example, instead of selling mass-produced inauthentic souvenirs, a museum could sell the works of local artists on its premises, with the profits being channelled to the artists. Similarly, hotels could sell local products at reasonable prices, with the profits being given to local entrepreneurs. In any case, organisations should inform tourists of their actions and be transparent about how they channel their profits. This may help them to be viewed by tourists as ethical (social) responsible entities.

(c) The third and most important layer is caring for others, including guests and employees. This is the core of the notion of *agape* in the context of tourism, and its deepest layer. An example is ensuring the physical and psychological comfort of the individual (whether guest or employee). It is offering something to the 'other' without expecting anything in return. Philanthropy and philanthropic actions coming from individuals or tourism organisations fall into this category. Care for the other often means 'putting the other person first', and it can take many and very simple forms. For example, someone in the line to pay for a trolley full of groceries might allow a person with a few items who is behind him/her in the queue to go first.

Christou (2018) found that when tourists referred to certain actions and behaviour by locals, hosts and tourism service providers, they perceived the following as expressions of *agape*: philoxenic attitudes, friendliness (from both service providers and locals), helpfulness and positive expressive behaviour. The last category included caring, pleasant and enthusiastic behaviour, a sincere smile, respectfulness and thoughtfulness, honesty, kindness, attentive high-quality service, politeness and a welcoming attitude. Examples given by tourists (Christou, 2018: 18) included the following:

> People should be welcoming and genuine towards everyone, and not be kind only to those that they like more, or have an interest in. I really like it when people ask me if I need help, or if there is anything that they can do for me, when I look distressed or worried. Sometimes it's good if people show that they care about you, without you necessarily saying to them that you need their assistance.
>
> Most people nowadays are 'independent,' they don't rely on others. Or, they act as if they don't need others ... Their tolerance level towards each other decreases day by day. And this is truly sad. But it's good if we ask for help. It's also good if we support each other too. This shows that we care about them, that we love them ... and not just ourselves.

In the above-mentioned study, pro-agapic actions were grouped as follows:

(a) genuine love for the other (*philalia*) in the form of philanthropic individual or organisational actions and comforting (physically and psychologically) those in need;
(b) genuine and warm-hearted hospitality in the form of thoughtfulness and kindness;
(c) care for the destination's environment (see Photo 2.2).

In contrast, actions undermining the notion of *agape* were classified as follows:

(a) love of one's self (*philaftia*) expressed through egoistic attitudes;
(b) impoliteness and inappropriate behaviour by managers, service providers or tourists;
(c) profiteering or profit-oriented actions by service providers or organisations such as restaurants, hotels and airlines;
(d) antisocial behaviour;
(e) lack of respect towards the natural and cultural environment of the destination.

It may be argued that actions that could be used to feed and cultivate personal and managerial *agape* involve a replacement of the 'I' mentality with a 'we' mentality and engaging in acts of kindness towards others, our colleagues and employees. These aims may be achieved through the cultivation of kind and positive thoughts about life and about others, based on the dynamic of thoughts in influencing and shaping relationships. As Thaddeus (2015: 69–70, 76) emphasised:

> Our life depends on the kind of thoughts that we cultivate. If our thoughts are peaceful and serene, if they have meekness and kindness, then this is how our life is ... If our thoughts are kind, peaceful and serene, then we influence our self and we radiate peace around us – in our family, in all of

Photo 2.2 Giraffe Centre, Kenya. The African Fund for Endangered Wildlife in Kenya created the Giraffe Centre with the aim of protecting this endangered species and releasing giraffes into safe game parks and conservation areas across the country. The Centre allows visitors to learn about endangered animals, engage with them actively and assist in their protection. These and similar initiatives may be regarded as pro-agapic actions towards the environment
Source: Author.

the country, everywhere ... When we cultivate negative thoughts, the evil is great. When there is evil inside us, we radiate it in our family and elsewhere we go. Catastrophic thoughts destroy our inner peace ... If you are peaceful and full of agape, if meek and kind thoughts monopolise in your mind, they [*people with negative thoughts*] will stop fighting you with their thoughts and they will cease to threaten you. But if you demand 'an eye for an eye,' then this means war. And where there is war peace cannot exist.

Lack of *agape* at a personal and managerial level may lead to a lack of communication between ourselves and others, and then to feelings of loneliness. According to Alexandrou (2014), a selfish person cannot communicate with others since they are egocentric and, essentially, unable to exit from their 'ego'. Eventually, the person becomes alienated and experiences loneliness. Papadopoulos (2014) noted the need to cultivate *agape* through philanthropy, particularly now, at a time when egoism, individualism, ruthlessness, greediness and materialism prevail.

Despite this analysis of the nature and importance of *agape*, it should be emphasised that this chapter does not suggest that prioritising service to others should be a person's ultimate goal. Unreasonable or immoral demands by guests cannot simply be accepted by hosts or employees for the sake of unconditional love. Likewise, the tensions that can arise between service to others and the need for self-development should not be underestimated. Moral issues and concerns in the tourism context are challenging, as tourism is a space where gratification, exploration and social engagement collide (see Caton, 2012).

3 Giving Philanthropy: Private and Organisational Philanthropy in Tourism and Hospitality

> There is your brother, naked, crying, and you stand there confused
> over the choice of an attractive floor covering.
> Aurelius Ambrosius, c.340–397

In its literal sense, philanthropy means being a friend of the human species, and it has been linked directly to 'goodness' and the habit of doing good (Bacon, 1985; Sulek, 2010). The notion has concerned philosophers throughout history (Machuca, 2006). There is evidence that philanthropy was practised by past societies, such as in Byzantium where it was viewed as a theological abstraction (an imitation of God's actions) and a political stance towards those in need (Constantelos, 2016). Philanthropy is founded on *agape*: it should be offered with no hidden agenda and without expecting any return in terms of personal or organisational gain (Christou, 2018; Christou & Sharpley, 2019; Papadopoulos, 2014).

According to Christou *et al.* (2019b), philanthropy should extend beyond the expression of sympathy to encompass actions that comfort other people. Such philanthropic actions may have direct recipients such as people in need, guests and employees, and indirect recipients such as a specific community or society as a whole. Philanthropic actions include the following: financial aid, the provision of space and philoxenia, provision of physical items and equipment, the offer of food and the provision of physical effort to actively help someone in need. The importance of philanthropy is acknowledged at societal, organisational and private levels. Woodruff (2018) argued that philanthropy can be a matter of life and death for institutions as well as for people, since human lives and quality of life often depend on philanthropy.

Private Philanthropy in the Tourism Context

> Good actions give strength to ourselves and inspire good actions in others. (Plato, *c*.428–*c*.348 BC)

There is evidence that people have a natural tendency to help those in need (Silk, 2006). In the wake of Hurricane Dorian (31 August–16 September 2019), hundreds of Airbnb hosts in regions of the south-east United States opened their homes to displaced neighbours and the relief workers that had been deployed to help. During that time, the Airbnb platform provided instructions to hosts on how to make their homes available for free (Airbnb, 2019). Studies have investigated various factors that affect people's willingness to participate in philanthropy (Henderson *et al.*, 2012), to show empathetic concern and to help those in need, such as personality characteristics (Bekkers, 2006) and level of education (Apinunmahakul & Devlin, 2008). Cultural norms and religious beliefs may also drive philanthropy, and philanthropy has been investigated in Jewish (Kosmin & Ritterband, 1991; Waxman, 2010), Muslim (Prihatna & Abidin, 2005), Christian (Constantelos, 1968; Vantsos & Kiroudi, 2007), Hindu (Osella *et al.*, 2015) and 'African' (Moyo & Ramsamy, 2014) contexts.

> Eliminate your sins with charity and your injustices with philanthropy to the poor. (John Chrysostom, *c*.349–*c*.407 CE)

Another reason for engaging in philanthropy is to gain personal ethical satisfaction. In the context of philanthropic travel, Katanich (2017) noted that tourists in this sector will return home with the best 'souvenir' – namely, the satisfaction of knowing that they have made a difference. As an informant in the study by Christou *et al.* (2019b) stated:

> Philanthropy is an essential part of our lives. What you give is what you'll get in life ... The joy of giving is great ... I experienced this when I gave to a family in need, and I saw the content they received and the surprise they had, since they weren't expecting it. The ethical pleasure of giving is great when you give and you can't easily express it through words. (Christou *et al.*, 2019b: 6)

In an extensive literature review of empirical studies of philanthropy, Bekkers and Wiepking (2011) identified eight mechanisms as the most important that drive charitable giving: awareness of need, solicitation, costs and benefits, altruism, reputation, psychological benefits, values and efficacy. The study by Christou *et al.* (2019b) on philanthropy within the tourism context revealed the following 'personal' motives for engaging in philanthropy:

(a) personal inner reasons ('because it feels nice');
(b) just reasons (the right for all people to live a decent life);

(c) past personal experiences (having been in a difficult situation);
(d) family reasons (having learned to offer by following the example of their parents);
(e) societal prospect reasons (in order to benefit society, channel goodness and avoid a future society with 'heartless' members);
(f) cultural and religious reasons (it is the right and moral thing to do);
(g) trust reasons (if philanthropy is offered at a specific organisation, then they trust the organisation to channel the money to those in need, and not to keep the charity money or any portion of it for their own interest).

Tourist philanthropy and 'voluntourism': Benefits, ethical issues and censure

While on holiday, tourists may also embrace philanthropy, with actions as simple as offering a poor child a chocolate bar, eating at a local restaurant in a remote area, buying crafts from local entrepreneurs, attending a philanthropic event, or joining a tour organised by homeless people. For example, the northern Italian city of Bologna was one of the first places in the world to offer tours guided by homeless people. The project, called 'Gira la cartolina' (flip the postcard), invites visitors to discover the other side of the city. One of the three protagonists behind this initiative, 'Daniele', is a former state employee with a university education who was evicted from her home and now lives in temporary accommodation provided by social services. Her experience in the theatre gave her the confidence to stand up in front of visitors and tell them about the history of the city's monuments (Fiorini & Amiel, 2019).

Tourists may also help the local community, for example by eating local food, not necessarily because they wish to engage in a sustainable/philanthropic travel experience but for their own economic reasons, that is, in order to save money. Under different circumstances, they might have opted for a beach vacation with an all-inclusive plan, or a luxury cruise with all meals and drinks included in the price.

> I have to admit I haven't really done well travelling in a sustainable manner since a months-long back-packing trip through Southeast Asia in my early 20s where cost, not commitment to a cause, had me eating local food, sleeping in train stations and taking cheap buses. (Glen, 2013)

Travel philanthropy has evolved through the democratisation of charity as a way of doing good by giving back while travelling (Novelli et al., 2015). Although people are generally less likely to help outgroup members (Levine & Thompson, 2004), individuals may travel to certain countries to offer their assistance to those in need, to help communities and their members and to share their skills and knowledge. Probably one of the earliest forms of travel philanthropy is the case of missionaries, individuals who choose to devote their whole lives to the well-being of others in different countries and settings as part of their mission. Such is the case of Fr Cosmas, as presented in Aslanidis and Grigoriatis (2001):

> Fr Cosmas loved the Africans very much and spent himself entirely in their service and for the improvement of their lives ... Whenever they learned that Fr Cosmas was around, they ran to meet him and seek a solution to their problems, whether these were of material or spiritual nature ... [He] would not eat, if all the Africans had not eaten first, because he was so fond of them. He served them with his own hands, and watched to see if they had any other needs when they had finished eating ... Apart from the poor and the sick, there were also the prisoners ... For those who had been imprisoned for minor offences, he would pay a sum for their release ... he would also give the price for a ticket home, or would take them in mission transport. (Aslanidis & Grigoriatis, 2001)

'Voluntourism' or volunteer tourism has been examined in a number of studies (Bandyopadhyay, 2018; Barbieri *et al.*, 2012; Bernstein & Woosnam, 2019; Coghlan, 2015a; Kontogeorgopoulos, 2017; Wearing & McGehee, 2013; Wong *et al.*, 2014). Tourism organisations and museums may actively seek volunteers, who are often asked to cover their own travel, accommodation and other expenses (including food). The official website of Auschwitz-Birkenau (2019) former Nazi concentration and extermination camp (and museum) asks for volunteer candidates such as adult students, journalists and teachers. Australian-based travel agencies market orphanage placements as 'voluntourism' – a blend of holiday and volunteering sold to well-intentioned travellers (Knaus, 2017).

Although volunteer tourism has become an established and commercialised market, Tomazos and Cooper (2012) have raised a number of ethical issues. While volunteer tourism meets the demands of the more morally conscious traveller, it primarily provides opportunities for economic gain for organisations that act as brokers for volunteer travel experiences. Nonetheless, volunteering may benefit the community and region in which services are offered, and it can also be a provider of gratifying personal experiences (Barbieri *et al.*, 2012; Strzelecka *et al.*, 2017). This is particularly true if voluntourism is undertaken for the right reasons, such as to offer help and serve others. Jake Dorothy volunteered to help in an elephant sanctuary in Thailand. There, he came across Kabu, a solitary elephant with a badly deformed front leg. The elephant was from the illegal logging trade, where it had been pulling huge weights on steep mountain ravines since infancy (Dorothy, 2016).

If run properly and ethically, orphanages may provide orphans with shelter, food, healthcare and education and help them build their future prospects. St Barnabas Orthodox Orphanage and School in Kenya was founded by Fr Methodius J.M. Kariuki and his wife Everlyn Mwangi. Today, the orphanage cares for around 180 children. A further 400 children have benefited from the project, and its sponsors/partners include the Ministry of Education of Kenya (St Barnabas Orthodox Mission Kenya, 2019). Even so, the nature of the need for orphanages and volunteers in such places has become a matter for debate. Riley (2016) argued that the use of volunteers is fuelling the growth of orphanages in Uganda, causing family members to become separated. This is because, in many cases, the orphanages are able to provide children with free education, meals and clothing and this encourages family separations. Knaus (2017) noted that Australians are among the top financial supporters of orphanages in a number of Southeast Asian countries such as Cambodia, but that orphanages may in fact exploit children for profit:

> Despite their good intentions, supporters of orphanages such as tourists and volunteers, actually end up contributing to the breaking up of families and removing children from their own family environment. (Morooka, in Knaus, 2017)

> I thought it [*the orphanage*] might be a good place. Maybe I could have enough food to eat, have a chance to go to school. But actually what I imagined is wrong ... He [*the director of the orphanage*] dressed us up looking poor so the visitors see us, they feel pity for us, and they donate more ... But they don't really know what was going on inside the orphanage. (Chan, a nine-year-old at an orphanage in Cambodia, quoted in Knaus, 2017)

Organisational Philanthropy in the Tourism Context

Organisational or corporate philanthropy concerns organisations donating a portion of their profits to non-profit organisations (Chen & Lin, 2015). Philanthropy is increasingly adopted by organisations and is playing a more prominent role in their visions and strategies (Hu & Yoshikawa, 2017; Schlegelmilch & Szőcs, 2017). Philanthropy may be linked with corporate social responsibility (CSR) activities (Su & Swanson, 2017), and Sheldon and Park (2011: 401) placed it in the category of 'philosophical and ethical issues'. Caroll (1991: 42) referred to it as the 'icing on the cake' of CSR: it is characterised by its voluntary nature, unlike CSR which is expected from organisations (Leisinger, 2007; Von Schnurbein *et al.*, 2016). As corporate philanthropy is not considered mandatory for businesses, it is regarded as an altruistic complement to CSR (Lantos, 2002) and may be abandoned in times of financial difficulty (Porter & Kramer, 2006). Thus, the scope of philanthropy is narrower than that of CSR (Farmaki, 2018; Farmaki & Farmakis, 2018).

There are various reasons for an organisation to engage in philanthropic actions, and the benefits for the organisation have been acknowledged in a number of studies (see Photo 3.1). Organisations may benefit by being perceived as socially responsible (Hallak *et al.*, 2012), by fostering relationships with the community (Christou *et al.*, 2019b), by building

Photo 3.1 Christmas charity trees at a resort in Orlando, USA. Unique personalised Christmas trees are decorated by local non-profit organisations and are displayed in the lobby area of a resort hotel in Orlando. These trees highlight the charitable missions and works of organisations throughout Orlando
Source: Author.

their reputation and financial capital(Brammer & Millington, 2005; Pan et al., 2018) and by increasing the engagement of their employees (Lee et al., 2014). To the extent that corporate philanthropy generates market competitiveness, tourism firms can use it as an effective strategy to compete for higher sales revenues and to achieve greater profits (Wang et al., 2018). Cruise firms may embrace philanthropic projects to justify their claims to be good corporate citizens and to mitigate censure of financial and social exclusionary behaviours such as using private enclave resorts as ports of call (Weeden, 2015). However, in a study by Christou et al. (2018a) of business motives for engaging in philanthropy, rural tourism stakeholders mentioned 'pressure' from charity organisations and from the community/society to provide help and support. Within the tourism sector, organisations such as airlines, hotels and tourist attractions may engage in philanthropic practices. With their campaign 'Shall We Make a Snowman with You', Turkish Airlines Volunteers distributed boots, coats, socks, blankets and clothes to almost 1700 children in villages across 19 provinces in Turkey (Turkish Airlines, 2019). In another example:

> For the third year in a row, the Emirates Airline Foundation funded a local charity in Niger to donate food supplies to poor families with young

children. The foundation donated USD 50,000 to Association d'Apui Pour les Services Humanitaires (AASH) to provide 650 baskets each with 50 kilos of rice, macaroni, powder milk, baby diapers and sweets. (Emirates, 2019b)

The study by Wang *et al.* (2017) concluded that tourism attraction companies that operate using public tourism resources and that have close relationships with neighbouring communities tend to engage more actively in corporate philanthropy than other companies in the industry. Hotels may combine charity with sustainable environmental and social practices. As Liu (2016) explained, some hotels work with charities to recycle used soaps left by guests in their rooms and donate the money to needy communities in Southeast Asia. In other cases, hotels hire people with specific financial, mental and physical difficulties as part of their philanthropic initiatives.

If you stay in one of the Lemon Tree hotels [*in India*], there's a chance the waiter serving you will be hearing-impaired or the receptionist could be an amputee. The direction of the hotel hires not only people with mental, or physical disabilities but also individuals with different emotional and financial needs. The hotels give a chance to those who may not have the easiest time in life. They make up, in total, one-fourth of the staff. (Katanich, 2017)

Criticisms of philanthropic initiatives: 'Hypocritical' and 'camouflaged' philanthropy

In response to the question 'What's worthy of philanthropy?', Sullivan (2019) reported on the backlash against the campaign for donations to rebuild Notre Dame Cathedral in Paris, one of the best-known tourist attractions and most recognisable buildings on earth, which went up in flames in 2019. The fire had hardly been put out when some of the richest people in France rushed to pledge their donations. This was followed by a protest by some of Paris's 3600 rough sleepers, asking how so many millions of euros could be found for a new cathedral roof but not a cent to put a roof over their heads (Chakrabortty, 2019).

As a Roman Catholic who has visited Paris many times, I was sad to learn that Notre Dame Cathedral was burning. But I became angry when I saw how quickly 'lovers of humanity' mobilized money for rebuilding a building in contrast to the sluggish pace at which institutional philanthropy moves money to human beings. (Lewis, 2019)

Ethical concerns are raised in cases where philanthropic initiatives and actions are perceived to be driven by insincere motives and egocentric reasons, that is, in order to satisfy someone's senses and curiosity or to fulfil experiential quests. Slum tourism involves visiting impoverished areas to experience another side of a destination (in some cases with the intention of helping to alleviate poverty in these areas). Mumbai's Dharavi slum, which occupies a plot half the size of Central Park in New York, is home to a million people. Local tour operators organise tours there. The slums are presented as a hive of industry, giving the impression of resourcefulness, despite the poor sanitation, lack of clean water, and squalid conditions. As Nisbett's (2017) study concluded, these slum tours enable wealthy middle-class Westerners to feel inspired, uplifted and enriched, with little understanding of the poverty there or of the need for change.

On another point, Bell (2014) noted that it has become commonplace to be asked to contribute to charity by sponsoring somebody's travel experience. Organisations arrange skydiving, mountain hikes, mountain climbing, exotic treks and other high-octane experiences, assuring participants that they will not be out of pocket if they put time and effort into collecting enough sponsorship.

> A friend asked for sponsorship to swim with sharks … as far as I'm concerned, that is asking me to pay for her hols … I found it too difficult not to. I gave her a fiver and I'm still annoyed with myself. But mainly annoyed with her. (Carter, in Bell, 2014)

Travel philanthropy may be combined with 'luxury' vacations. In such cases, tourists stay in upscale hotels or boutique establishments and enjoy luxurious experiences while supporting local people. Although this can be seen as a 'win-win' situation for tourists and locals, the travel philanthropy and luxury concept raises the question of whether this act is actually an expression of 'private' philanthropy; the rewards and benefits extend to the individual tourist, who in addition to having a rewarding personal experience enjoys a luxurious stay and lavish gastronomic experiences. Katanich (2017) noted the availability of luxurious resort holidays in a hotel in Nicaragua which was created by the best-known philanthropists in the country to help the local community. The luxury hotel's presidential suite boasts a personal concierge and a park ranger to guide tourists in their reconnection with nature. In a similar case in South America, an antique Tren Ecuador luxury train takes visitors to heritage sites with local guides and artisan markets, thereby sustaining thousands of jobs for people living in remote communities.

Tourism organisations may be subject to criticism for their choice of philanthropic actions, which may be perceived by some people as unethical, dishonest, deceitful or simply not serving the true cause of philanthropy. Ryanair introduced a lottery game in which some of the profits are donated to charity. The airline claimed that, over a period of five years, sales of these scratch cards generated over 2 million euros for good causes, including Childline and SOS Children's Villages. However, the game has been criticised on the grounds that only a small percentage of the revenue actually reaches good causes, and that players have only an estimated one in 1.2 billion chance of winning the million-euro jackpot. This is because the winner is invited to a room where 125 envelopes are laid out across a table, just one of which contains the top prize (Gillespie, 2016). On the basis of an internal staff memo, Power (2017) reported that the airline's cabin crew are required to sell a specific number of scratch cards each per day or face action. A spokesman for the airline said that 'for commercial reasons' they could not disclose the total value of scratch cards sold in a year.

Christou *et al.* (2019b) suggested that organisations should be transparent about the extent to which they offer philanthropy. They provided the following taxonomy of people's perceptions of private and organisational philanthropy:

(a) *Ultimate philanthropy*, where philanthropic actions by individuals or organisations are perceived by people as honest and genuine and as having no direct benefit to the provider. For instance, if a person sacrifices his/her personal interest without waiting to be asked to offer assistance, this could be perceived as honest philanthropy.
(b) *Critical philanthropy*, where philanthropic actions are questioned for their genuine merits. In such cases, perceptions are blurred, since it is unclear whether the actions are driven by true interest towards others or by other insincere personal/organisational interests. It may be argued that perceptions of corporate philanthropy are 'critical', since people may understand that philanthropy is underpinned by reasons linked to the promotion, image and branding of the organisation.
(c) *Hypocritical philanthropy*, where philanthropic actions are founded on hypocrisy. In such cases, individuals and organisations 'camouflage' their true nature under philanthropic actions so that they are perceived as ethical, anthropocentric and considerate towards those in need. Yet, both recipients and the general public may easily understand that these actions are not sincere or genuine. For example, an organisation's philanthropic actions may be perceived as phony if it is seen as not treating its employees in ways that are fair, just and ethical:

> [Organizations] must firstly respect their fellow human being-employee and then do 'hypocritical' philanthropy to others. If you do not truly love the person who works for you even for years, then how can you call yourself a philanthropist or, an organization based on philanthropic values? (Research respondent, quoted in Christou *et al.*, 2019b: 8)

In this taxonomy, the outcomes at a personal, organisational and societal level can be either positive or negative. A negative personal outcome of practising 'ultimate philanthropy' is for someone to be deprived of money and free time and to be burdened with the effort of helping others. A positive outcome, which could outweigh the negatives, is the rewarding personal experience and ethical pleasure of giving to someone in need. For an organisation, a negative boomerang outcome could be damaging to the business's image if its philanthropic actions are perceived as 'hypocritical'.

Similarly, at the societal level, if philanthropy is not targeted properly, it may lead to unequal and unjust distribution of resources.

Conceptualising Philanthropy in the Tourism and Hospitality Context

In the previous sections of this chapter we have discussed motives for engaging in private and corporate philanthropy, types of philanthropic actions, criticism of corporate tourism philanthropy, and how people perceive the philanthropic actions of individuals and of organisations. These drivers, acts and perceptions enable the construction of a useful diagram which illustrates how philanthropy is manifested within the context of tourism (see Figure 3.1). In the diagram, the column on the left shows the drivers of both private and organisational philanthropy. As mentioned previously, drivers of private philanthropy include religious beliefs and the personal fulfilment derived from engaging in philanthropic actions; drivers of organisational philanthropy include pressure from society and from charitable organisations to donate money, products and services.

In the case of small businesses, private and organisational drivers seem to overlap. There is room for debate as to which personal drivers are manifestations of *agape*, but the act of engaging in philanthropy with the aim of gaining something in return is surely a non-agapic philanthropic motive. Likewise, it can be argued that corporate drivers of philanthropy are not founded on *agape* because of the benefits resulting from philanthropy, as is the case when philanthropy is offered with the aim of improving the company's image. These drivers are translated into actions. Hence, an act of private philanthropy can be a donation of money or an offer of psychological support to an individual in need. Travelling for volunteer purposes may be regarded as a form of philanthropy, provided that the volunteer offers his assistance for the benefit of others and not for his/her own interests. Tourists may also engage in slum tourism as a way to help relieve poverty. However, it is arguable whether this is an expression of philanthropy, as tourists' motives are often linked with experiential pursuits and curiosity. Acts of organisational philanthropy include donations to charity organisations. All these private and corporate actions may influence perceptions of the philanthropy as 'ultimate', 'critical' or 'hypocritical' (Christou *et al.*, 2019b).

As Figure 3.1 shows, perceptions of ultimate philanthropy are created when people perceive philanthropic actions as honest without a hint of personal or organisational interest. On the other side of the spectrum, perceptions of hypocritical philanthropy are created when people believe that individuals or organisations are engaging in philanthropic actions as a way to increase their financial status or improve their public image. In this case, the organisations do not really care about humans in need or the environment they live in, and their actions, such as mistreatment or taking

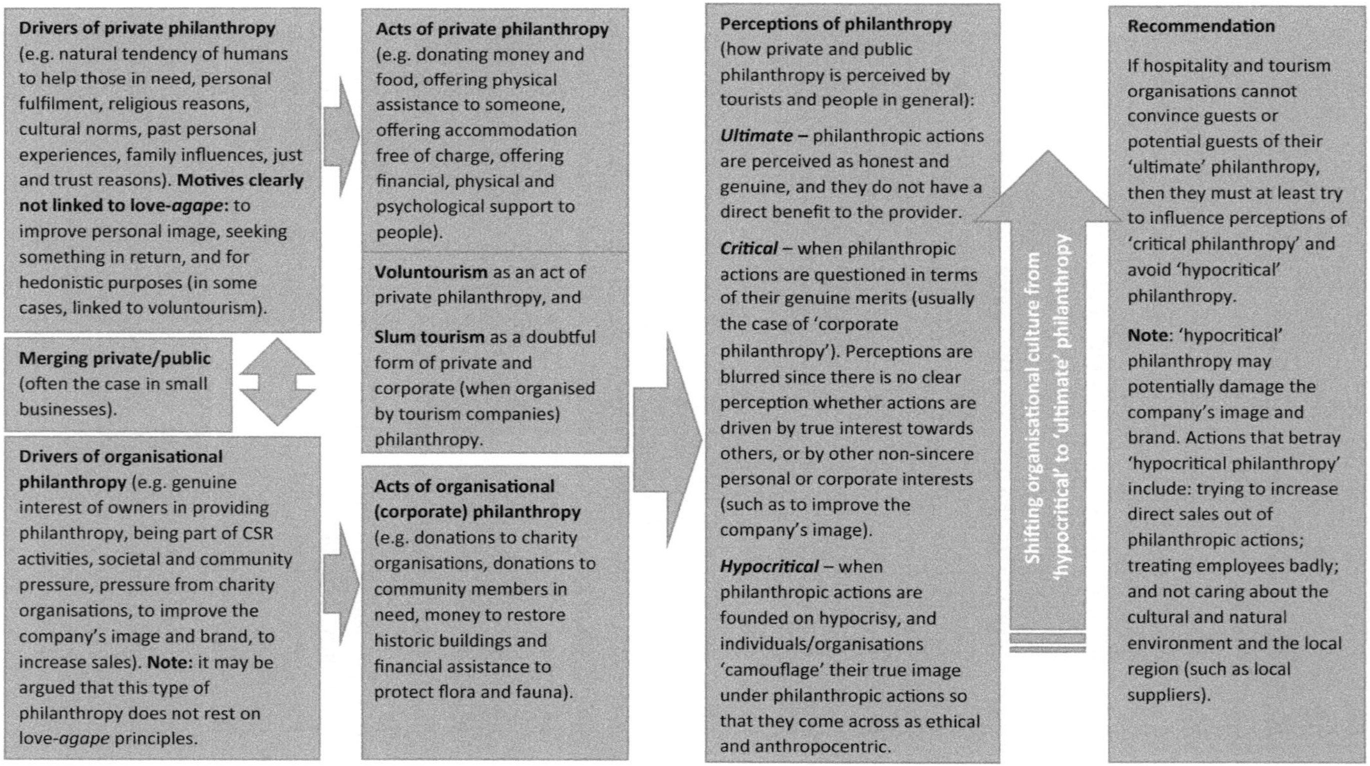

54 Part A: The Philosophy of *Giving* in Hospitality and Tourism

Drivers of private philanthropy (e.g. natural tendency of humans to help those in need, personal fulfilment, religious reasons, cultural norms, past personal experiences, family influences, just and trust reasons). **Motives clearly not linked to love-*agape*:** to improve personal image, seeking something in return, and for hedonistic purposes (in some cases, linked to voluntourism).

Acts of private philanthropy (e.g. donating money and food, offering physical assistance to someone, offering accommodation free of charge, offering financial, physical and psychological support to people).

Voluntourism as an act of private philanthropy, and

Slum tourism as a doubtful form of private and corporate (when organised by tourism companies) philanthropy.

Perceptions of philanthropy (how private and public philanthropy is perceived by tourists and people in general):

Ultimate – philanthropic actions are perceived as honest and genuine, and they do not have a direct benefit to the provider.

Critical – when philanthropic actions are questioned in terms of their genuine merits (usually the case of 'corporate philanthropy'). Perceptions are blurred since there is no clear perception whether actions are driven by true interest towards others, or by other non-sincere personal or corporate interests (such as to improve the company's image).

Hypocritical – when philanthropic actions are founded on hypocrisy, and individuals/organisations 'camouflage' their true image under philanthropic actions so that they come across as ethical and anthropocentric.

Recommendation

If hospitality and tourism organisations cannot convince guests or potential guests of their 'ultimate' philanthropy, then they must at least try to influence perceptions of 'critical philanthropy' and avoid 'hypocritical' philanthropy.

Note: 'hypocritical' philanthropy may potentially damage the company's image and brand. Actions that betray 'hypocritical philanthropy' include: trying to increase direct sales out of philanthropic actions; treating employees badly; and not caring about the cultural and natural environment and the local region (such as local suppliers).

Shifting organisational culture from 'hypocritical' to 'ultimate' philanthropy

Merging private/public (often the case in small businesses).

Drivers of organisational philanthropy (e.g. genuine interest of owners in providing philanthropy, being part of CSR activities, societal and community pressure, pressure from charity organisations, to improve the company's image and brand, to increase sales). **Note:** it may be argued that this type of philanthropy does not rest on love-*agape* principles.

Acts of organisational (corporate) philanthropy (e.g. donations to charity organisations, donations to community members in need, money to restore historic buildings and financial assistance to protect flora and fauna).

Figure 3.1 Philanthropy in tourism and hospitality
Source: Author.

advantage of employees, make this clear. As a result, the organisation's philanthropic actions are viewed as 'hypocritical', and its image may be damaged as a result.

Therefore, it is recommended that hospitality and tourism organisations act (both internally and externally) in such a manner that they create a perception of ultimate philanthropy. Care for employees, the community and local people is as important as efforts to help people in need in disadvantaged countries or regions, as an organisation cannot be called truly anthropocentric or philanthropic if it does not care for the well-being of its employees both inside and outside their working environment. Specific suggestions for organisations acquiring a philanthropic culture are summarised below.

(i) Philanthropical support within an organisation

First, an organisation should try to encourage an organisational culture built on respect and care. It should assist any of its members who are in a problematic situation or experiencing difficulties within or outside their workplace. Furthermore, organisations should not only treat their employees fairly and with respect but should also try to identify employees who are in specific need of extra financial support (e.g. for medical reasons). The organisation should help these people discreetly, without asking other employees to contribute. If colleagues wish to contribute, then they must do so without being 'forced' to by the organisation.

(ii) Philanthropical support at the local community and national levels

It is recommended that any hospitality and tourism organisation should engage in philanthropic actions towards charity organisations or people in need within the community they operate in (see Photo 3.2). In addition to support for the local community, it is recommended that the organisation engages in philanthropic actions towards charitable bodies that operate at a national level.

(iii) Philanthropical support at an international level

There is an ongoing debate about which charity organisations, especially in disadvantaged countries, support people in real need, such as orphans. Owners of companies, or key people in managerial positions, should actively seek to identify non-profit charitable organisations that operate in an ethical manner, genuinely helping people in need and not taking advantage of them or benefiting from the money raised. Organisations may also consider the formation of a volunteer group (including employees at all levels) to travel to disadvantaged regions to provide financial and

Photo 3.2 Bran Castle, Transylvania, Romania. Tourism and hospitality organisations can channel benefits towards the local community by supporting locally produced crafts, helping local entrepreneurs, buying from local suppliers and offering employment opportunities to disadvantaged people. Here, during the Easter season at the famous Bran (Dracula's) Castle in Transylvania, Romania, artworkers carefully paint and decorate hollow eggs in a traditional design
Source: Author.

physical aid and assistance to specific groups of people and communities. This may offer employees a unique experience outside the workplace and opportunities to help/feel valuable, to foster teamwork dynamics and perhaps to build loyalty towards the organisation. In some cases, it is enough to ask people to provide help, as they are ready and willing to support a good cause. In Ethiopia in 2019, millions of people across the country were invited to take part in a major initiative to plant trees; in only 12 hours, 353 million trees were planted (Paget & Regan, 2019).

Important note: Any 'promotions' of philanthropic organisational initiatives must be done in a discreet manner. For instance, instead of advertising an initiative in a local newspaper, it should be included subtly on the organisation's official website. Although the exposure of philanthropic actions may improve a company's image and strengthen its brand name, not doing so may avoid its being criticised for 'hypocritical' and 'critical' philanthropy.

Academics have emphasised the need for further research to understand and appreciate philanthropy at both the personal and corporate levels (Christou *et al.*, 2019b; Von Schnurbein *et al.*, 2016; Zhang *et al.*, 2016).

Part B

The Philosophy of *Receiving* in Hospitality and Tourism

4 Receiving Experiences: Tourist Senses and Emotions

The Role of the Senses and Hedonism in Tourism

A number of studies have investigated the sensory dimension of tourist experiences (Agapito *et al.*, 2014; Kim & Fesenmaier, 2015). In particular, the significant role of the senses in influencing travel motivation, informing tourism marketing, shaping tourist experiences and affecting visitor satisfaction has been acknowledged in a number of studies. According to Cohen and Cohen (2019), most sensory tourism research has embraced the conventional division and hierarchy of the senses attributed to Aristotle (with sight being in first place, followed by hearing, smell, taste and touch). The importance of the senses has been highlighted in specific forms of tourism, including virtual experiences in thematic tourism (Martins *et al.*, 2017) and theme hotels/restaurants (see Photo 4.1), and also in food, wine and gastronomy tourism (López-Guzmán & Sánchez-Cañizares, 2012). Disney's imagineers understood that scent is a strong trigger of memory, and scent-producing machines placed in various locations all over Disney parks overwhelm visitors with mouth-watering aromas of freshly baked cookies and popcorn (Fantozzi, 2018; Lennon, 2019). Conversely, travel and tourism organisations may prohibit certain actions, elements and circumstances that impact the senses negatively. For example, in September 2019 Japan Airlines announced a new booking tool for travellers who want to avoid being seated near crying infants on flights (Wallace, 2019).

Tourists may be attracted to specific regions or even particular hotels in order to enjoy particular tastes or to drink specific beverages. An example is the famous Raffles Hotel Long Bar in Singapore, where the iconic pink-hued 'Singapore Sling' cocktail was created in 1915 by bartender Ngiam Tong Boon and continues to attract guests from around the world (De Jong, 2018). The 'Philly Cheesesteak' is of one of the most famous things to come out of the Pennsylvanian city of Philadelphia. Today the cheesesteak is an attraction in its own right, and visitors can sample it at

Photo 4.1 Venice resort, Macau. The re-creation of particular places is common in the tourism industry, as in this representation of parts of Venice at a resort hotel in Macau. The inclusion of sale points and commercial stands may increase money inflows for the organisation, yet may impact negatively on the aesthetics of the set; the commercial presence is in tension with the as-close-as-possible 'authentic experience' that the resort appears to have been aiming to create
Source: Author.

multiple locations (Euronews, 2019). Another example is the famous 'Sachertorte' chocolate cake, which was invented in Austria in 1832. Visitors queue in front of the café of the Sacher Hotel in the centre of Vienna to buy the 'Original Sacher Torte'.

Malone *et al.* (2014) noted that the focus of much research on the sensorial aspects of the tourist experience has been on the 'gaze', that is, on a single sense. Nonetheless, Isacsson *et al.* (2009) argued that in order to influence tourist motivation and tourist consumption, all the senses must be involved, not only the visual sense. As part of their promotional efforts, destinations and tourism organisations can use comments linked to human senses, or elements that may stimulate the human senses of their potential visitors:

> Alpine Summer is a feeling that every visitor can experience with all senses. The smell of sun-warmed hay during mowing time, while hiking towards the top through freshly mown meadows and watching farmers rhythmically swing their scythes, is surely unforgettable. The whooshing sound of the scythe blades can be heard from afar, just like the cheerful yodelling one might hear echoing across the valley. A rest along the way is definitely worth it! Nowhere else can you view the majestic mountain panorama better than on a soft, mossy pasture. Here, between blooming blueberry shrubs, the gaze goes on forever and satisfaction and tranquility slowly set in. (Salzburgerland, region in Austria, 2019a)

> Kyoto City attracts millions of local and international visitors each year looking for traditional Japanese culture … Stay in a traditional ryokan, take a dip in a rejuvenating onsen, and enjoy the seasonal changes of cherry blossoms and brilliant autumn foliage. (Japan National Tourism Organization, 2019)

> Each time I travel to New Zealand, I find new culinary delights, more delicious wines and more often than not a few taste bud surprises. (M. Swaine, culinary travel writer, quoted on the official site for Tourism New Zealand, 2019)

'Attention to detail', as a way to stimulate the senses and provide a sense-filling experience, has always been the focus for the luxurious travel and hospitality industry (see Photo 4.2). Upscale airlines and luxury hotels try to impress their guests with fine details and extraordinary features. For instance, in certain hotels, floor lighting activates as soon as the guest steps out of bed, lighting the way to the bathroom.

> Great luxury hotels since the late 1800s have focused on innovation and tiny details. Cesar Ritz, considered by many the grandfather of the modern luxury hotel, was the first to introduce a bathroom in every room in 1893 at the Grand Hotel in Rome. Today, you wouldn't imagine a hotel room without a bathroom, but back then it was a small detail that changed the industry. (Hirschowitz, president of the International Luxury Hotel Association, quoted in Krueger, 2019)

Christou *et al.* (2018a) proposed the acronym HASE (holistic agreeable senses experience) as an important part of the satisfaction process of

Photo 4.2 Smurf Village, Belgium. Attention to detail may trigger visitors' senses, impacting favourably on the tourist experience by creating feelings of 'cuteness', as in this colourful miniature display at Smurf Village in Belgium
Source: Author.

visitors (see Photo 4.3). In particular, the authors argued that the offerings of tourism service providers should satisfy all the human senses in an agreeable and pleasant manner, through the absence of endogenous and exogenous incidences or factors that might disturb this process (that is, hazardous conditions or unruly behaviour by hosts or tourists). As an example, they considered the organisation of a 'cultural event'; according to the HASE approach, the event must include the following elements:

(i) aesthetics: traditional, well-preserved buildings and surroundings;
(ii) entertaining cultural music;
(iii) distinct pleasant smells (from endemic flora and traditional food);
(iv) fresh and tasty local food for visitors to experience;
(v) the opportunity for visitors to touch and/or be actively engaged (for example, by dancing) in the event process.

Furthermore, the 'tangibility' of the tourist experience may well be enhanced through the provision of souvenirs. These play a threefold role: (a) tangibilising the tourist experience; (b) contributing to site branding (if carefully chosen and well designed); and (c) acting as memorabilia for the tourist experience (Christou & Farmaki, 2019).

Photo 4.3 Seaplane Harbour, Estonian Maritime Museum, Estonia. Many museums have been keen to offer their visitors a holistic sensual experience, transforming spaces of 'object display' into interactive and educational centres through innovative use of technology, interactive displays and memorabilia fostering the branding of the museum. The Seaplane Harbour in Tallinn has strong links with Estonia's heritage and provides a holistic visitor experience
Source: Author.

Pertinent in connection with the senses and how to please them is the notion of 'hedonism'. The term derives from the Greek word for delight (*hedonismos*). Democritus (*c.*460–*c.*370 BC) was the first philosopher to embrace a hedonistic philosophy, in which he referred to 'contentment' as the supreme goal of life. The philosophical school of the Cyrenaics, founded in the 4th century BC, taught that the only intrinsic good is pleasure, in the sense of positively enjoyable momentary sensations and not simply the absence of pain. In Hebrew, one possible interpretation of 'Eden', the name of the garden where God placed Adam and Eve, is 'pleasure'. In the tourism context, hedonism has been found to influence satisfaction evaluations (Rodríguez-Campo *et al.*, 2019), visitors' behavioural intentions (Coudounaris & Sthapit, 2017) and tourists' subjective well-being (Sthapit & Coudounaris, 2018). Lonely Planet (2012) described 'hedonistic city breaks' (in Berlin, Ibiza, Las Vegas, Tel Aviv, Budapest and Istanbul) in terms of serious partying, shopping, fine dining and/or massages.

Although hedonism is arguably linked with pleasure derived from the senses, it has also been associated with 'thrilling' and 'exciting' experiences (Wang, M. *et al.*, 2019). 'Alternative hedonism' is emerging as individuals dissatisfied with consumerism and materialism seek the 'good life' (Soper *et al.*, 2009). From many spiritual and ethical perspectives, individuals are urged to be very cautious as to how they use and attempt to please their senses, since sensory pleasure may lead to negative consequences for oneself and others through intoxication, adultery, lewdness and gluttony (Cavarnos, 1994).

> The carnal person believes that the only enjoyment is that of the senses so [s]he does not dominate over them ... but allows [her]himself behave like an irrational animal and gets carried away by [her]his senses: [S]He runs ... to enjoy all lawlessness, [s]he seeks pleasure as a purpose and considers it honorable, even if [s]he finds it in the greatest of filth and uncleanness. (Nicodemus the Hagiorite, 1749–1809)

Pertinent here is the case of the sexual exploitation of children in tourism (travelling to engage in sexual acts with minors), a phenomenon that plagues nations worldwide (Newman *et al.*, 2011). Using India as a case study, Chemin and Mbiekop (2015) argued that public investment in child protection is necessary to shield children from potential sex tourists. Malone *et al.* (2014) claimed that interpretations of ethical tourism should be revised to take account of hedonism in ethical choices and ethical tourism experiences. In regard to hedonism and seeking to satisfy and please the senses, the following quote taken from Farasiotis (2005) is thought provoking:

> He [*the typical contemporary person*] doesn't want to struggle to go beneath life's superficiality, because modern culture has made him comfortable as he lives the pampered life of a hungry consumer in a cage of materialism. In a state of spiritual death, his life is defined by the biological life of his body, and his interests are defined by his exclusive fixation

on his bodily needs and desires ... When man despises the immortal soul, the higher part of his being, and becomes deeply attached to his body, he becomes utterly carnal and condemns himself to a life that is degrading. (Farasiotis, 2005)

Triggering tourists' senses: The paradigm of 'nostalgia'

The notion or phenomenon of nostalgia has been subject to critique when it is based in racism or other cultural and political issues (see, for example, Barnett, 1982; Brooker, 1987; Dann, 1995; Goldstein & Hall, 2017). There has been discussion of 'pathological nostalgia', a longing for the past without accepting that is over (Kaplan, 1987), and in many cases nostalgia may be an unsatisfactory or even harmful experience (Verplanken, 2012). Nevertheless, nostalgia has recently shifted from being an indicator of psychopathology to being seen as something more positive (Smeekes, 2015; Specht & Kreiger, 2016), especially within the tourism context (Christou, 2020a; Christou *et al.*, 2018a) (see Photo 4.4).

A number of studies have recognised that nostalgia potentiates an attainable future, since it generates self-positivity and optimism for the future, increases perceptions of life as meaningful and fosters pro-social behaviour (Cheung *et al.*, 2016; Sedikides & Wildschut, 2016; Sedikides *et al.*, 2015). A variety of (in)tangible elements may trigger nostalgia in visitors and guests,

Photo 4.4 Sovereign Hill, Ballarat, Australia. An open-air thematic museum, such as this one in the city of Ballarat, Australia, does not necessarily trigger feelings of 'personal nostalgia'. However, it may create nostalgic feelings for a country's past heritage and an era that could only be experienced today with the help of a time machine

Source: Author.

from a particular place and the structures and elements within it, such as a beach (Jarratt & Gammon, 2016), to a region with the traditional architecture, objects, smells and nostalgic experiences that it offers to visitors (Christou *et al.*, 2018a). Photographs and pictures (Kessous & Roux, 2008; Pan, 2011), music, songs and specific sounds (Chou & Lien, 2010), sharing old stories with others (Davalos *et al.*, 2015) and specific tastes such as traditional food (Vignolles & Pitchon, 2014) can also trigger nostalgia. Christou's (2020a) conceptual study attempted to link the construct of nostalgia with tourism, revealing certain (in)tangible elements that address the senses and that may act as triggers of nostalgia:

(a) the ambience of a place (for example, the exterior and interior design of a building), which addresses the visual sense;
(b) the consumptive element (that is, specific food, such as traditional food, or food offered at specific social gatherings), which addresses both the visual and the non-visual senses;
(c) the olfactory element (specific scents that may evoke memories of the past), which addresses the non-visual senses;
(d) the acoustic element (for instance, nostalgic music, as well as the exchange of old stories in social gatherings), which addresses the non-visual senses;
(e) the constructive environment and its elements (artificial structures, pictures, objects and memorabilia), which addresses both the visual and the non-visual senses;
(f) the natural environment of the destination (its specific/distinct landscape and endemic flora and fauna), which addresses both the visual and the non-visual senses.

Emotions in Tourism

Daily I expect murder, fraud, or captivity, but I fear none of these things because of the promise of heaven. (Pádraig of Ireland, *c*.386–461)

The study of emotions

Research interest in emotion has grown enormously (Ekman, 2016) in a range of social disciplines, including tourism (Gao & Kerstetter, 2018; Lin *et al.*, 2014; Mackenzie & Kerr, 2013). This increased interest derives from the fact that emotions are reliable predictors of human behaviour (Gaur *et al.*, 2014; Malone *et al.*, 2014). Moreover, emotions have been linked to tourist experiences (Faullant *et al.*, 2011), satisfaction (Hosany *et al.*, 2016), organisational success and branding (Brotheridge & Lee, 2008; Rossiter & Bellman, 2012). An impressive number of studies have investigated emotions within the context of tourism, and tourism studies have equipped us with the knowledge necessary to understand the generation, regulation, manifestation and dynamics of emotions, particularly

Photo 4.5 Landscape, Iceland. Specific landscapes, dramatic terrains, imposing mountains and impressive waterfalls, such as this one in Iceland, may evoke 'awe' in visitors. It has been noted that people can form a deep connection to places and may feel a specific form of 'love' for them, often referred to as 'topophilia'
Source: Author.

within the experiential milieu (Gao & Kerstetter, 2018; Kim & Fesenmaier, 2014; McCabe & Johnson, 2013) (see Photo 4.5).

Nevertheless, the characterisation of 'emotion' is challenging (Izard, 1978), and emotion is a difficult phenomenon to study (Gaur *et al.*, 2014). For instance, fear and anger are accepted as emotions, but it is arguable whether surprise is an emotion (Power & Dalgleish, 2008). Izard (1971) identified ten basic emotions that can blend together to give different feelings. Ekman *et al.* (1972) found evidence for six basic emotions. Power and Dalgleish (2008) proposed five basic emotions, including 'happiness'. Ekman (1982) drew up a list of 17 emotions, and Lazarus (1991) also proposed a long list, including anger, anxiety, fright, guilt, shame, sadness, love and hope. A recent study by Ekman (2016) revealed a high degree of agreement in the scientific community that there are at least five emotions: anger, fear, disgust, sadness and happiness. The emotions of 'shame', 'surprise' and 'embarrassment' were endorsed by approximately half of the participants of Ekman's (2016) study. Other emotions, such as 'guilt', 'contempt', 'awe', 'envy' and 'love' attracted substantially less support.

In a study by Christou (2018), where respondents were asked to give examples of emotions, 'love' was found to be the emotion most commonly referred to, with almost half of the respondents naming it. Emotions can also be described on different levels, including the philosophical, the psychological and the sociological. For instance, researchers (Yeh *et al.*, 2012) referred to 'nostalgia' as an emotion. Rahmani *et al.* (2018) made reference to the emotions of 'trust' and 'anticipation' within the context of tourism. Moreover, cultural views and dynamics should also be acknowledged. For example, Hindus regard 'heroism' as an emotion (Hejmadi *et al.*, 2000), and Japanese regard *amae*, the pleasant feeling of depending on someone else, as an emotion (Niiya *et al.*, 2006).

Despite the ongoing discourse on emotions, Izard (1978) found general agreement that a very broad classification of emotions into positive and negative is generally correct and useful. However, Izard went further, categorising emotions in terms of their intra-individual nature, or as relations between the person and the environment (for example, anger may be correlated with survival). Existing systems for the classification of emotion seem to agree on one fact: that there are more negative emotions than positive ones (Ekman, 1982; Izard, 1971; Zautra, 2003). Each negative emotion has a distinct facial expression, whereas positive emotions differ relatively little from one another in expression (Ekman, 1982). For instance, a genuine smile is a good indicator of someone's true feelings, given that it is hard to produce one voluntarily (Kalat, 2011).

Emotions seem to prepare the body for very different kinds of responses. For instance, with anger, a rush of hormones provides energy for action (Goleman, 2005). Strong emotions also make people's breathing faster (Gomez *et al.*, 2005). When we are frightened we inhale deeply, and if we are disgusted we partly close our eyes (Susskind *et al.*, 2008). Emotions are our primary motivation system (Izard, 1978), and they prompt us to explore new opportunities (Fredrickson, 2001). They also focus our attention on important information and images; for example, we turn towards a pleasant image or away from an unpleasant one, even if we are trying to pay attention to something else (Schupp *et al.*, 2007), and we remember information better if it is emotionally arousing (Levine & Pizarro, 2004). Emotions also regulate our priorities; if we see something frightening, for example, we concentrate on the danger as if we could not see anything else (Mathews & Mackintosh, 2004). Researchers have provided evidence in support of certain theories of the emotions, how they are expressed, and why and how they are generated. The key classical theories of emotion are summarised in Table 4.1.

Academic research on tourism and emotions

Prima facie, emotions are important within the tourism context. They affect the tourist experience (Christou, 2018a) and have therefore been used to understand the behaviour of consumers (Kozub *et al.*, 2014). Significantly, they have been acknowledged as a key mediating factor in the guest satisfaction process (Io, 2016) and in retention/loyalty crescendos (Han *et al.*, 2009; Lo *et al.*, 2015). Certain streams of research have focused on the investigation of emotions within the context of tourism (see Table 4.2):

Table 4.1 Classical theories of emotion

Theory	Author/s	Source/s	Key assumption/main theme
Evolutionary theory	Charles Darwin	Ekman, 1973	Emotions evolved because they were adaptive, allowing humans/animals to survive and reproduce (e.g. love leads to reproduction and fear to either fighting or fleeing the source of danger).
James–Lange (or Thalamic) theory	William James, Carl Lange	James, 1884	Emotions occur as a result of physiological reactions to events ('I am trembling because I saw a lion. Therefore I am afraid').
Cannon–Bard theory	Walter Cannon, Philip Bard	Cannon, 1927	We feel emotions and experience physiological reactions (e.g. trembling) simultaneously. This theory contradicts the James–Lange theory.
Schachter–Singer (or Two-Factor) theory	Stanley Schachter, Jerome Singer	Schachter & Singer, 1962	A stimulus leads to a physiological response, which is then cognitively interpreted and labelled and results in an emotion. The theory draws on both the James–Lange and Cannon–Bard theories.
Cognitive Appraisal (or Lazarus) theory	Richard Lazarus	Lazarus, 1991	Thinking must occur before emotion is experienced. A stimulus followed by a thought leads to the (simultaneous) experience of a physiological response and an emotion (e.g. I see a lion, and I think that I am in great danger; this leads to fear and the physical reactions associated with the 'fight-or-flight' response).
Facial Feedback theory	Charles Darwin, William James	Ekman, 1973; James, 1884, 1894	Facial expressions are linked to the experience of emotions. Emotions are directly tied to changes in facial muscles. Physiological responses are not only a consequence of emotion but also have a direct impact on emotions (e.g. people who are forced to smile pleasantly at an event, instead of maintaining a neutral facial expression, will have a better time).

(1) A first key stream of research has examined the results and impacts of tourist activity and tourism development on the emotional states of visitors and residents (Jordan *et al.*, 2019; Shakeela & Weaver, 2012).

(2) A second research stream has produced a significant number of studies related to tourism experiences and their impacts on the emotions of travellers, visitors and guests. Specifically, emotion regulation has been examined in the context of holidays (Gao & Kerstetter, 2018) and of tourist motivation and behaviour (Yan *et al.*, 2018). Rahmani *et al.*'s (2018) study revealed that anticipation and trust are important tourism drivers, and in the context of experiential tourism the hospitable attitudes of locals, hosts and employees towards guests have an important role to play. The significance of hospitality gestures has

Table 4.2 Key themes of tourism research related to emotions

Tourism key theme linked to emotions	Paradigm of tourism themes linked to emotions	Exemplar	Relevant studies
Tourism development and impacts	Tourism impacts, development, incidents and emotions (such as reactions of tourists and residents to impacts)	How tourism activity and tourism development impact on the emotional state of visitors and residents.	Jordan et al., 2019; Shakeela & Weaver, 2012; Zheng et al., 2019
Tourism experiences	Emotional challenges and emotion regulation (while on holiday)	Psychological intervention of tourists to maximise the positive emotional outcomes of their own experiences.	Gao & Kerstetter, 2018; Sedgley et al., 2017
	Tourism experiences and emotions (such as while on vacation)	How tourists' experiences shape/affect their emotional states. The role of emotions within an experiential tourism context. The role of emotions in creating extraordinary and memorable experiences.	Christou, 2018; Kim & Fesenmaier, 2014; Lin et al., 2014; Mitas et al., 2012; Moal-Ulvoas, 2017; Podoshen et al., 2015; Rahmani et al., 2018
	Tourist behaviours, intentions, motivation, hedonism, satisfaction, well-being, perceived brand relationship and attachment, value creation and emotional solidarity	The interrelation of emotions with tourist satisfaction. Emotions (e.g. trust) that influence tourist motivation. Emotions that shape the future intentions of tourists. The role of emotions in brand interaction and relationship. Emotions and customer value creation. Emotional solidarity (emotional bonds created between tourists and residents/hosts).	Rodriquez del Bosque & San Martin, 2008; Faullant et al., 2011; Hosany, 2011; Hosany & Gilbert, 2009; Io, 2016; Malone et al., 2017; Prayag et al., 2015; Woosnam, 2012; Woosnam & Norman, 2009; Yan et al., 2018
Forms of tourism and emotions	Adventure tourism, ethical tourism, dark tourism and rural tourism, and impacts of these (and other forms of tourism) on the emotional states of visitors	The impact of a particular form of tourism on the emotional state of tourists. Engagement in specific tourism-related activity and how this affects visitors. Forms of tourism disclosing a strong emotional component (such as spiritual tourism and dark tourism).	Buda et al., 2014; Malone et al., 2014; Podoshen et al., 2015
Places and settings	Relationship of places and emotions (such as familiar places and rural settings)	Emotional bonding with particular settings (topophilia). The impact of settings on the emotional state of tourists (e.g. nostalgia).	Christou, 2020b; Hosany et al., 2016; Pardwardhan et al., 2019; Pearce, 2012

(Continued)

Table 4.2 *(Continued)*

Tourism key theme linked to emotions	Paradigm of tourism themes linked to emotions	Exemplar	Relevant studies
Internet places/social platforms	The role of social media in creating or influencing the emotional states of visitors and guests	The effects of social media on emotions.	Hudson et al., 2015; Jabreel et al., 2017
Tourism workplace and organisations	Tourist entrepreneurs and employee workplaces and emotions (e.g. tour guides and hospitality employees)	Emotional intelligence in the workplace. The case of emotional labour in the work environment. Emotions and emotional expressions of service providers towards guests.	Mackenzie & Kerr, 2013; Van Dijk et al., 2011; Wong & Wang, 2009

been highlighted in a number of studies (see Lashley, 2015a) which draw primarily on the influence of hospitableness on the emotional states of guests, as well as emotional relationships between hosts/service providers and guests. Lugosi *et al.* (2016) found that gestures of hospitality can generate emotions that are positive, thus building loyalty. Christou (2018) stressed the significance of *agape* (a sincere form of love) in building strong and heartfelt relationships between hosts and guests. Another important concept is philoxenia, where hospitality is provided by people with no expectation of anything in return and logic is replaced by wholehearted love towards the guest (Mantzarides, 2005). In the study by Christou *et al.* (2018a), informants made reference to positive emotions and words such as love, empathy, warmth and comfort, after being offered and having experienced philoxenia. This was the result of a simple action, such as the offering of a cold beverage on a hot day without being charged.

(3) A third stream of research has investigated particular forms of tourism and the impact of participation and engagement on the emotional states of tourists (Malone *et al.*, 2014).
(4) A fourth stream of research has looked at the influences of places on visitors' emotional states (Hosany *et al.*, 2016; Pearce, 2012), including topophilia (Tuan, 1974), consumption tourist sites (Knudsen & Waade, 2010) and the evocation of emotions that are not necessarily positive (Christou, 2020b), as in the case of dystopian dark tourism and thanato-places (Podoshen *et al.*, 2015).
(5) A fifth stream of research has examined emotions and their importance and regulation within the tourism workplace. This has produced important insights into the now well-established notion of 'emotional

labour' (whether at the surface level or deep acting) and how emotions are expressed (Van Dijk *et al.*, 2011) in organisations. Moreover, the emotional skills of managers and employees are increasingly recognised as essential for everyday workplace and managerial practices (Brotheridge & Lee, 2008). Relevant in this case is emotional intelligence (EI), the importance of which has been highlighted in a number of studies (Riggio & Reichard, 2008; Tsaur & Ku, 2017). Research has also suggested that managers with high levels of EI may use such skills to promote their own agenda, hence raising ethical concerns. For instance, an individual may control emotions to facilitate the accomplishment of different personal goals (Kilduff *et al.*, 2010).

Human interactions and spaces influencing the emotional states of tourists

Hitherto, research on the nexus of tourism and emotion has provided significant insight into how the emotional states of tourists and people who interact with them, such as hosts and service providers, are shaped through their interaction in particular physical settings. Studies have investigated the case of Oahu, Hawaii (Jordan *et al.*, 2019) and the banks of the River Thames in the UK countryside. Dynamics incorporating human transactions and spaces (such as settings and regions with intense/low tourist activity) have an impact on the emotional states of tourists as well as locals, as illustrated in Figure 4.1.

The matrix is informed by the review of the literature above. The vertical axis of the matrix represents the intensity of human interaction between tourists and locals/service providers and with other tourists. The horizontal axis gives the physical settings in which these human interactions take place, such as a hotel or, on a larger scale, an island. The matrix illustrates the interrelation between these two dynamics and the impact that these may have on emotional states (including those of visitors). Studies have also investigated the impacts of tourism in regions at mature phases of the destination life cycle (Vargas-Sanchez *et al.*, 2015). Doxey (1975) proposed the causation theory, according to which a gradually increasing number of tourists in particular settings may cause local people to become irritated and may build on mutual negative emotional states that result from irritations expressed overtly by both parties.

A recent study (Zhang *et al.*, 2018) exposed particular spaces within cities that experience crowds because of an increased number of visitors. These experiences impact on the tolerance levels of residents towards visitors, with some residents making hostile and angry statements such as 'they [visitors] are not welcome at all'. Nevertheless, settings and spaces that experience low levels of touristic activity, or that are characterised by relatively small amounts of human interaction between tourists and

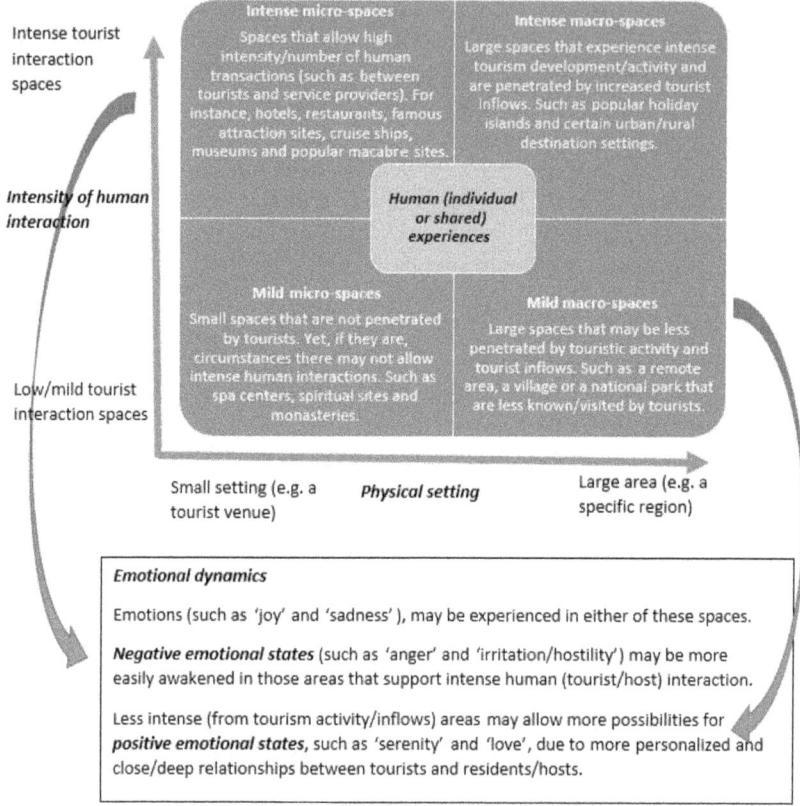

Figure 4.1 Intensity of human interactions and spaces influencing emotional states
Source: Author.

service providers/locals, may foster more positive emotional states. Moufakkir and Selmi (2018) found that visitors experienced positive 'deep emotions' in a large desert area. Other recent studies have supported the causal effect of settings and spaces that accommodate low tourism activity/flows of tourists on visitors' emotional states, such as generating the emotion of 'love'. This finding may be partially explained by opportunities in these settings for hospitality offerings that are warmer and more personalised (Christou & Sharpley, 2019).

Conceivably, close/intimate deep psychological interfaces between hosts and guests or emotional solidarity dynamics (Woosnam, 2012; Woosnam & Norman, 2009) may be supported in cases that allow personal relationships and common experiences to be built. However, one should not come to the false conclusion that positive emotional states are not generated within settings that support high levels of tourism activity/interaction, such as hotel settings (Mody et al., 2019) or vice versa (see Photo 4.6). It should also be borne in mind that tourists may apply emotion regulation strategies during their experiences (Gao & Kerstetter,

Photo 4.6 Island of Ithaca, Greece. Tourists may experience a sensual gastronomic experience in a beautiful setting, with pleasing aesthetic surroundings, enjoyable music and delicious food. However, if they experience negative emotions such as anger and sadness as a result of impolite behaviour by service providers or profiteering conduct by venue owners, their experience is less likely to be fulfilling. Here, restaurants are set in the eye-pleasing and picturesque 'mythical' island of Ithaki, Greece
Source: Author.

2018). Another important factor in visitors' emotional states is the type of setting involved. Natural (that is, physical) settings appear to have greater impacts, according to studies of a mountain setting (Pomfret, 2006) and of the 'utopic' natural setting of Iceland (Christou & Farmaki, 2019).

The tourism and emotions nexus

A fortiori, the role of positive emotions in the experiential transaction between service provider and guest has received much attention in studies.

> Tourists prefer to be assisted and served by friendly and smiley people … I prefer that my hosts express to me their emotions if they are positive, to feel more comfortable and happy. I prefer not, if they are negative. (Tourist, quoted in Christou *et al.*, 2019a: 156)

Research and evidence derived from studies related to the emotion and tourism nexus enable the construction of a diagram presenting the main pillars of the elements closely associated with emotions within the tourism context (see Figure 4.2). The conceptual diagram summarises the key elements and dynamics involved in the generation, regulation and expression of emotions within the tourism context.

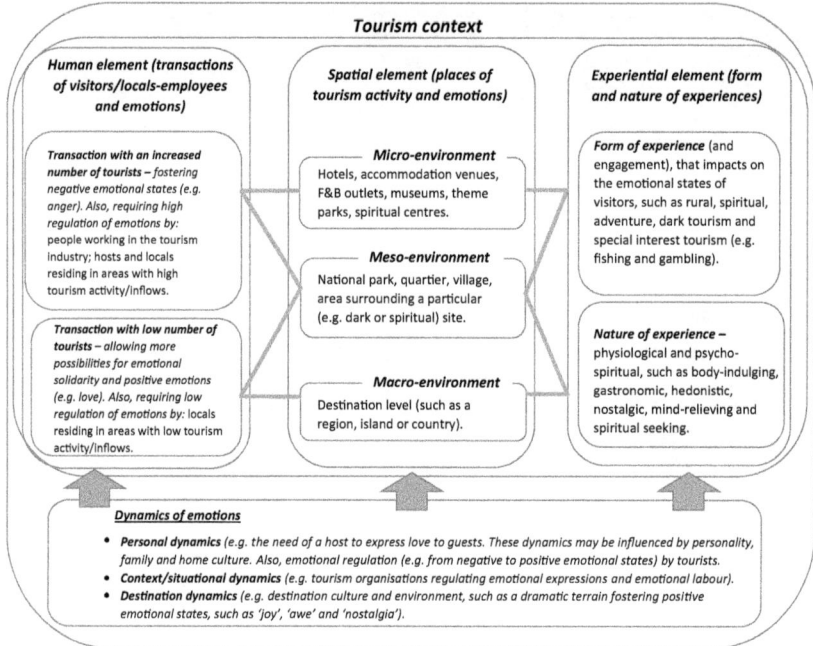

Figure 4.2 Conceptualising the tourism and emotions nexus
Source: Author.

(a) The first key element deals with human transactions (for example, between tourists and service providers) and the emotional dimensions that such transactions entail. This element involves the connection of tourism development, impacts and emotions (Jordan *et al.*, 2019; Zheng *et al.*, 2019) and the (low or high) intensity of human transactions as a result of tourism development and tourist inflows. Jordan *et al.* (2019) noted that tourism impacts significantly on the emotions of people residing in areas where tourist activity takes place. Likewise, tourist experiences are affected by the closeness between tourists and hosts (Woosnam, 2012).

(b) The second key pillar element is the spatial area in which emotional transactions take place: at the micro level a hotel (Lo *et al.*, 2015) and at the macro level a destination. It should be emphasised that emotional transactions also take place within the internet space. Social media platforms provide many opportunities for visitors to exchange their opinions and emotions (such as 'love', 'joy', 'sadness' and 'anger') to tourism service providers and others through words, comments or emojis. Parkwell (2019) has shown how emojis have emerged as an important feature of modern discourse through the spread of social media, which is commonly used by tourists for the purpose of complaining (Dolan *et al.*, 2019).

(c) The third key pillar is the experiential element. This dimension involves the form and nature of the tourist experience and the impact of this experience on the tourist's emotional state (Malone *et al.*, 2014; Pratt *et al.*, 2019; Sharpley & Jepson, 2011). The tourism and hospitality industry may shape the nature of a tourist's experience in a positive way by 'going the extra mile', targeting the psychological needs of guests and offering 'unexpected experiences', as in the following example:

> We [*at Hyatt Regency Delhi*] received a call from a guest saying he had accidentally left behind his child's favourite doll and that it had greatly upset the child … we couriered the toy to his home in London, but before that, we took photographs of the doll being served food, of its sleeping and watching TV, and mailed the images to the guest so his child would know the doll was being taken care of and would feel happy. (Executive Housekeeper Namrata Menon, in CN Traveller, 2015b)

These three key pillars are interrelated because human, spatial and experiential elements intermingle, as in the case of human interactions between a host and a guest taking place in the spatial margins of a village. Certain dynamics are involved in the process of generation, regulation and expression of emotion: personal dynamics such as personality influences (Faullant *et al.*, 2011); organisational and situational dynamics such as emotional labour (Van Dijk *et al.*, 2011); and destination dynamics involving place (setting) and culture. For instance, the physical environment or morphology of a destination may create emotions of 'awe' and 'joy'.

Tourism planners and regulators ought to consider and acknowledge the effects of tourism activity on the emotional states of both tourists and residents, especially in spatial settings that support intense human transactions, as is the case with special events and areas close to popular tourist sites. Service providers should also consider which precise emotions they wish to evoke in their guests/visitors: 'joy', 'nostalgia' or even 'sadness'? Christou *et al.* (2018b) argued that certain emotions that are negative by valence (for instance, 'fear' and 'sadness') may nevertheless have a positive effect on the experience of visitors, especially in certain theme parks or in dark tourism sites. The same authors found that causing 'love', 'nostalgia', 'enthusiasm', 'happiness', 'joy' and 'pride' may have an affective outcome and lead to a satisfying visitor experience. On the contrary, causes of 'frustration', 'disappointment', 'anger' and 'boredom' will impact unfavourably on the visitors' experience overall.

This chapter has built on existing theoretical knowledge of emotions within the tourism context, and in doing so has identified opportunities for further research. First, it would be interesting and helpful for tourism stakeholders if tourism academics examined specific emotions that may have escaped the attention of researchers so far (e.g. Hope). The tourism context at the micro and macro levels may provide direction for such research, and a phenomenological approach would be particularly suitable for these exploratory studies. Secondly, another stream of research could investigate the neglected role of technology in influencing and shaping emotional relationships between visitors and service providers, organisations and destinations as a whole. Thirdly, although it is clear that the role of emotions in dystopian places, such as in dark tourism, is complex (Podoshen *et al.*, 2015), it would be helpful for visitors if academics investigated the extent to which positive emotions (such as 'hope' and 'nostalgia') can be promoted in places where tragedy and death have occurred (Stone & Sharpley, 2008).

Fourthly, research could usefully focus on findings from potential studies linked to the emotional reactions of people in contexts that support intense human transactions and in settings that necessitate low interaction between hosts and tourists. Some studies have already examined how tourism impacts people's emotional states (Jordan *et al.*, 2019). The academic community could build on this by identifying thresholds of tourist activity and mass tourism concentration which, if exceeded, may lead to negative emotional states for both residents and tourists.

Fifthly, it would be interesting to consider how positive emotions may be generated and how emotional solidarity (as in Woosnam, 2012) may be promoted within contexts and situations that require very low, limited or no physical interaction with tourists. A particularly interesting and relevant case would be social media platforms (Hudson *et al.*, 2015; Jabreel *et al.*, 2017). Finally, a number of studies have examined how forms and types of tourism activity/engagement impact on the emotional states of tourists (Lee & Kyle, 2011; Malone *et al.*, 2014; Nawijn *et al.*, 2012). It would be worth examining the extent to which visitors' emotional states generated by tourism (from gambling to spiritual tourism) are transmitted and thus affect the emotional states of their partners, friends and colleagues.

As a final note, it is difficult to disagree with Sedgley *et al.*'s (2017: 22) statement that 'tourism experiences are journeys of mixed emotions'. Tourism experiences affect emotions, and vice versa. No doubt investigations in the area will continue to increase for this reason.

5 Receiving Happiness: Tourist Well-being and Psychology

> Stand at the brink of despair, and when you see that you cannot bear it anymore, draw back a little, and have a cup of tea.
> Sophrony of Essex, England, 1896–1993

The Notion of Well-being

'The only thing scholars working with the concept of wellness, more commonly referred to as well-being, seem to agree on is that there is no agreed upon definition of it' (Caton *et al.*, 2018: Introduction). Nevertheless, well-being (also referred to as wellness) can be characterised broadly as the condition of an individual or group, with a high level of well-being implying that the individual's or group's condition is positive.

> It [well-being] is an affirmation of your own being and of existence as a whole. It is a feast of the senses, an abundance of liveliness, an explosion of color, a belief of 'I am and I can' and 'we are and we can.' (Munar, 2018: xiii)

The concept of well-being evolved around two philosophies:

(i) the 'hedonistic' view, based on the 4th century Greek philosopher Aristippus, who considered that the goal of life should be to avoid pain and to experience as much pleasure as possible (Carlisle *et al.*, 2009).
(ii) the 'eudaimonistic' view. The term *eudaimonia* derives from the Greek words for 'good' (or well-being) and 'spirit' (or minor deity, that is, someone's controlling destiny). The notion relates to the realisation of human potential through a focus on well-being connected to meaningful and valuable actions, as opposed to 'vulgar' pleasure seeking (Boniwell, 2008). It is linked to self-development and optimal performance of meaningful behaviour (Cloninger, 2004). 'Hedonistic' pleasure-seeking activity provides instant well-being; in contrast, 'eudaimonic' effects can result even from activities that are unpleasant at the time but have delayed positive effects (see Smith & Diekmann,

2017). For example, beauty spas may offer 'hedonistic' experiences, whereas spiritual retreats offer experiences that are more 'eudaimonistic' (Voigt et al., 2011).

The increased study of well-being across several disciplines reflects a growing recognition of its value across the major domains of life (Bartels, 2015), including the field of tourism (Grimwood et al., 2018) (see Photo 5.1). Well-being is difficult to study empirically (Kringelbach & Berridge, 2017), and researchers tend to disagree on exactly what it constitutes (Goodman et al., 2018). More specifically, the causes of subjective well-being (SWB, a self-reported measure of well-being, typically obtained by questionnaires) may differ, including between age groups (Twenge et al., 2016). Likewise, cultures differ in their levels of well-being and the types of SWB they value most (Diener et al., 2018). Cooke et al. (2016) identified 42 instruments that varied significantly in their length, psychometric properties and operationalisation of well-being. They concluded that there remains considerable disagreement regarding how to understand and measure well-being.

Despite this disagreement, measuring the well-being of citizens has become an established practice in many advanced nations (Austin, 2016). The Organisation for Economic Cooperation and Development (OECD) Better Life Index uses the following dimensions to gauge living conditions and quality of life, with Sweden being the top performer for well-being (Hutt, 2019):

(a) subjective well-being (life satisfaction);
(b) income and wealth (such as household income);
(c) jobs and earnings (such as employment and earnings);
(d) housing (such as housing affordability and rooms per person);
(e) work–life balance (working hours and time off);
(f) health status (life expectancy and perceived health);
(g) education and skills (such as educational attainment and adult skills);
(h) social connections (social support);
(i) civic engagement and governance (such as having a say in government);
(j) environmental quality (air quality and water quality);
(k) personal security (such as feeling safe at night).

According to Ryff (1989), there are three sub-disciplines in psychology regarded as critical for the study of psychological well-being: developmental psychology, personality psychology and clinical psychology. (It may be argued that the absence of mental illness contributes to someone's well-being.) Psychological well-being and SWB have been closely linked to happiness (Diener et al., 2002, 2018; Stone & Mackie, 2013; Zhang et al., 2017). Walsh et al. (2018) mentioned that researchers use the term SWB when considering happiness, although SWB may include other measures such as those that tap negative emotions. Higher SWB has been associated

Photo 5.1 Prague, Czech Republic. The organisation by destinations and communities of seasonal and aesthetically pleasing markets and fairs, such as this one during the Easter period in Prague (Czech Republic), may contribute to some extent to general well-being. This is because they can be generators of positive feelings among locals, visitors and local entrepreneurs through the creation of appealing and unusual surroundings, pleasing music and increased opportunities for socialising, buying and consuming local products
Source: Author.

with good health, better social relationships, good work performance and creativity (Diener *et al.*, 2018). Particular activities and life events may have an impact on people's SWB; for instance, Kim and James (2019) found that participation in sport has a positive relationship with short- and long-term SWB. Life events, such as divorce, death of a spouse or unemployment, are also associated with lasting changes in SWB (Lucas, 2007). Figure 5.1 summarises the main points related to the philosophical and psychological perspectives of well-being discussed above.

The Notion of Happiness

> Happiness resides not in processions, and not in gold, happiness dwells in the soul. (Democritus, *c*.460–*c*.370 BC)

> Man only likes to count his troubles; he doesn't calculate his happiness. (Fyodor Dostoevsky, Russia, 1821–1881)

A 'happy' person has been defined by a number of researchers as one who frequently experiences positive emotions such as joy, happiness and contentment (Boehm & Lyubomirsky, 2008; Walsh *et al.*, 2018) (see Photo 5.2). From a positive psychology approach, happiness is a product

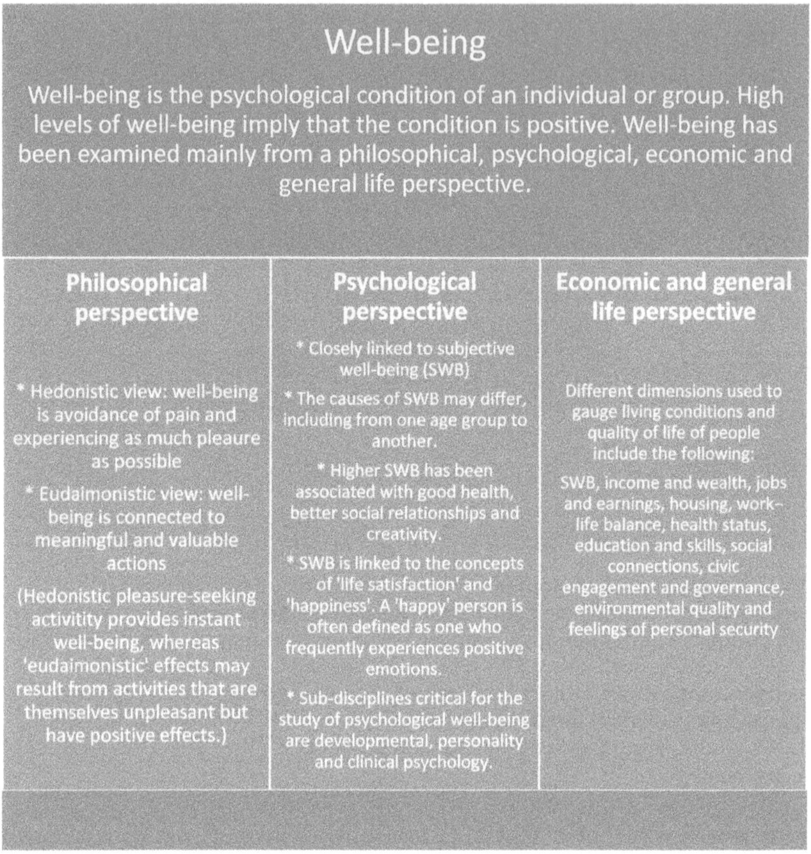

Well-being

Well-being is the psychological condition of an individual or group. High levels of well-being imply that the condition is positive. Well-being has been examined mainly from a philosophical, psychological, economic and general life perspective.

Philosophical perspective	Psychological perspective	Economic and general life perspective
* Hedonistic view: well-being is avoidance of pain and experiencing as much pleasure as possible * Eudaimonistic view: well-being is connected to meaningful and valuable actions (Hedonistic pleasure-seeking activity provides instant well-being, whereas 'eudaimonistic' effects may result from activities that are themselves unpleasant but have positive effects.)	* Closely linked to subjective well-being (SWB) * The causes of SWB may differ, including from one age group to another. * Higher SWB has been associated with good health, better social relationships and creativity. * SWB is linked to the concepts of 'life satisfaction' and 'happiness'. A 'happy' person is often defined as one who frequently experiences positive emotions. * Sub-disciplines critical for the study of psychological well-being are developmental, personality and clinical psychology.	Different dimensions used to gauge living conditions and quality of life of people include the following: SWB, income and wealth, jobs and earnings, housing, work–life balance, health status, education and skills, social connections, civic engagement and governance, environmental quality and feelings of personal security

Figure 5.1 Well-being and perspectives
Source: Author.

of positive emotions, engagement and meaning (Filep, 2008). According to Lyubomirsky *et al.* (2005), the main components of happiness are a frequent positive affect, high life satisfaction and infrequent negative affect. Nevertheless, it is acknowledged that happiness is a somewhat complex concept, going beyond positive, subjective experiences and perceptions to include certain ethical and moral dimensions (Smith & Diekmann, 2017). In fact, viewed through a religious prism, happiness is still founded on and depends on ethics and the ethical means by which it is achieved (Khan, 2019).

The philosophy of happiness

The philosophy of happiness is mentioned here in conjunction with ethics and morality, since it was associated with the notion of being well, but also with the notion of doing well. In the *Nicomachean Ethics*,

Photo 5.2 Mini-Europe theme park, Belgium. Theme parks, amusement parks and water parks may make visitors smile, laugh and be thrilled, ultimately triggering the emotion of 'joy'. Yet, it is unclear whether this response can be classified as 'happiness'. These 'joyful' places may channel some of their profits into charitable activities, so that visitors may feel that they are contributing to the well-being of disadvantaged people, and not just to their own. The photo shows a theme park in Belgium where visitors can feel like 'giants' by strolling through miniature replicas of famous buildings of Europe
Source: Author.

Aristotle postulated that happiness is the only thing that humans desire for their own sake, unlike riches or friendship. For the philosopher, the term 'eudaimonia' (broadly translated as 'flourishing and happiness') is more an activity than a state or emotion. Aristotle understood that the happy life is a good life in which someone fulfils human nature in an excellent way. For Aristotle, performing one's function excellently and rightly is good, and true happiness is found by leading a virtuous life (Smith & Diekmann, 2017).

Later, drawing on the work of thinkers ranging from Epicurus (Greek philosopher, 341–270 BC) to Jeremy Bentham (English philosopher, 1748–1832), a family of consequentialist ethical theories known as utilitarianism developed. These theories promoted actions that maximise the happiness and well-being of the majority of a population. Unlike egoism, utilitarianism considers the interests of all beings equally (Lazari-Radek & Singer, 2017). Utilitarian philosophers agree with Aristotle that subjective feelings of happiness are not the ultimate goal (Ryff & Singer, 2008), but they go on to maintain that the greatest number of people should derive happiness from a morally good action (Bentham, 1789; Mill, 1998 [1863]). Arguably, with the rise of individualism, the links between duty and happiness have gradually broken, with happiness being tied less closely to social life and social well-being and directed instead towards individual psychology.

'Studying' happiness

Happiness has been given extensive attention by researchers. It is often viewed as an important or even pre-eminent life goal, and is associated with positive outcomes at the personal, organisational and societal levels (Gentzler *et al.*, 2019; Zwolinski, 2019). From a review of the academic literature, Walsh *et al.* (2018) concluded that happiness is correlated with and often precedes career success, and that experimentally enhancing positive emotions leads to improved outcomes in the workplace.

Although happiness has been widely studied by philosophers and psychologists, it is only more recently that it has attracted the attention of academics from other disciplines, including tourism (Bimonte & Faralla, 2012). Happiness is often considered subjective, yet there are many similarities to be found in the reasons that underpin people's happiness (Nawijn & Peeters, 2010). Participants in the study by Bojanowska and Zalewska (2016) were asked to list associations that came to mind on hearing the word 'happiness' and they associated the word most often with good health and relationships. Other categories included knowledge, work, material goods and freedom. The Subjective Happiness Scale (SHS), which consists of four items measured on a 7-point scale, has been widely used to measure happiness (for example, Nawijn & Peeters, 2010).

What contributes to (un)happiness?

According to Lyubomirsky *et al.* (2005), around 50% of the differences in people's happiness levels are accounted for by their genetically determined set points. Another 40% of the variance may be explained by intentional activity and the remaining 10% by unintentional activity. According to the 'Easterlin paradox', formulated in 1974 by Richard Easterlin, a professor of economics, happiness varies directly with income among and within nations, but happiness does not trend upward as income continues to grow. The results of Zhang *et al.* (2017) also supported the conclusion that economic growth may not bring more happiness.

Other sociopolitical and employment factors may play a role in someone's happiness. For instance, the study by Pacek *et al.* (2019) found that in industrial democracies public employees are 'happier' and exhibit greater life satisfaction than people who are in otherwise similar circumstances. In addition, physical activity has been associated with well-being and happiness (Khazaee-Pool *et al.*, 2015). According to Quoidbach *et al.* (2019), there is strong evidence that engaging in social relationships promotes happiness. People report feeling happier when they are with friends and family members than when they are alone (Reis *et al.*, 2000; Sandstrom & Dunn, 2014). There is also a positive correlation between how happy people generally feel and the amount of time they spend with family and friends (Diener & Seligman, 2002), and this is supported by substantive conversations rather than small talk (Mehl *et al.*, 2010).

However, a number of factors seem to have an adverse effect on happiness, such as poor health, environmental reasons and income inequality. Oishi *et al.* (2011) found that an increase in income inequality may lead to unhappiness. Zhang *et al.* (2017) found that air pollution reduces hedonic happiness and increases the rate of depressive symptoms. Liu *et al.* (2016) found that in middle-aged women poor health can cause unhappiness.

Drawing on these findings about the properties of happiness, Figure 5.2 summarises some key ways in which happiness is understood.

The 'transfer' of happiness from one person to another

Humans communicate happiness in ways such as smiling and hugging. Transmission of emotional states may occur by different modalities (visual, auditory and olfactory) through their respective manifestations (Semin, 2007; Semin & de Groot, 2013). Feelings of happiness transfer between individuals through mimicry induced by vision and hearing. De Groot *et al.* (2015) observed that exposure to body odour collected from senders of chemosignals (odours produced by the body) in a happy state induced a facial expression and perceptual-processing style indicative of

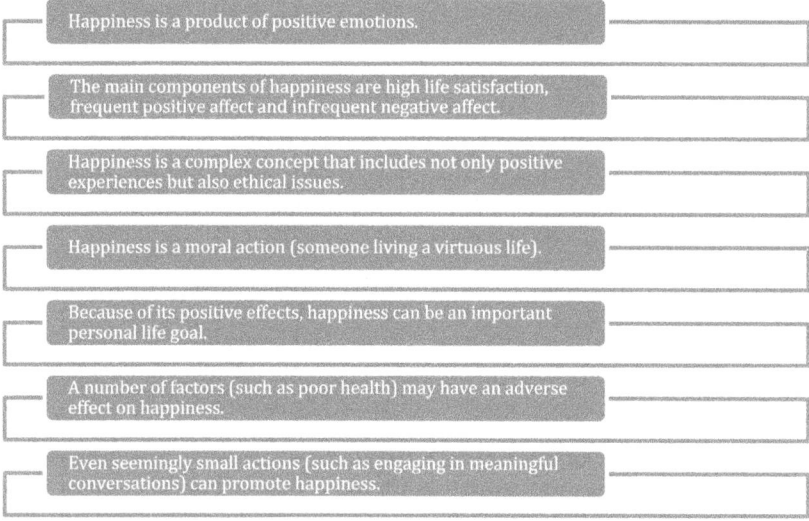

Figure 5.2 Ways of understanding happiness
Source: Author.

happiness in the receivers of the signals. Thus, they concluded that not only negative affect but also a positive state, such as happiness, can be transferred by means of odours.

In pursuit of happiness? An ethical and spiritual perspective

> Do not spoil what you have by desiring what you have not. Remember that what you now have was once among the things you only hoped for. (Epicurus, *c*.341–270 BC)

People may pursue happiness, and it may seem obvious that there is nothing intrinsically wrong with that. Nevertheless, the pursuit of happiness can, paradoxically, impair well-being (Ford *et al*., 2015). Research with adults suggests that placing an excessive value on happiness (for example, worrying about happiness even when one feels happy) is linked to lower SWB and higher incidence of depressive symptoms. Likewise, young people who value happiness too much are found to be more likely to be depressed (Gentzler *et al*., 2019). However, the pursuit of happiness for the common good can benefit both the individual and society, thereby reinforcing the very foundation of 'happiness': the link between happiness and ethics, being well and doing well, as discussed above. As Ford *et al*. (2015) reported, in collectivistic cultures, the pursuit of happiness may be more successful (compared to individualistic cultures) because happiness is viewed and pursued in ways that are relatively socially engaged.

From a spiritual perspective, happiness rests in our hearts and can be the result of someone's connection with God, not the result of physical acquisition or the pursuit of happiness in hedonistic actions. This kind of 'connection' will ensure that a person finds peace in their heart and mind, feels content and spreads joy to others through acts of kindness. For instance, consider the words below (derived from Kivotos, 2019) of St Nektarios (1846–1920), famous for his philosophical and theological views and his academic work and ministry in a number of cities, including Alexandria, Athens, Cairo, Chios and Istanbul (and whose life and teachings are available in a number of books: Chondropoulos, 1997; St Nektarios, 2016; St Nektarios of Aegina, 2019):

> How much are people deceived when they seek happiness away from themselves, to foreign countries and travel, to wealth and glory, to great fortunes and pleasures and to all the vile and inferiorities that end in bitterness! The erection of the tower of happiness outside our hearts is like a building being built on a land savaged by constant earthquakes. Soon such a building will land on earth … Happiness is within yourself, and happy is the person who understood it. Examine your heart and see its mental state. Did she lose her desire for God? … Does your conscience accuse you of wrongdoing, lying, neglecting your duties to God and people? Inquire if malice and passion filled your heart … Unfortunately, the one who neglected his heart deprived all goods and fell into a

multitude of evils. He took away the joy and filled it with bitterness, sadness and sorrow. He drove away peace and gained anxiety, turmoil and terror. He banished love and accepted hatred. (Derived from Kivotos, 2019)

Well-being and Happiness in the Context of Tourism

The art of living happily is to live in present. (Pythagoras *c.*570–*c.*495 BC)

Tourism as a contributor to happiness and well-being

Vacations spent in a different environment have been found to help people detach psychologically from work and other mundane concerns (Chen *et al.*, 2013); in one example, following a ski vacation, mood and life satisfaction levels were improved and mental stress levels were alleviated. However, planning a trip may increase stress linked to route planning and coordinating work tasks for the period of absence (De Bloom *et al.*, 2010).

It is recognised that tourism is very much a business of attempting to fulfil idealised desires (Smith & Diekmann, 2017). A number of studies have identified positive connections between happiness and holiday trips (Gilbert & Abdullah, 2004; Kemp *et al.*, 2008; Nawijn & Peeters, 2010; Neal, 2000). Engaging in certain activities and actions while on holiday may have a positive effect on people's happiness. Bimonte and Faralla (2012) recorded higher levels of happiness for participants who engaged in tourist activities, and Gillet *et al.* (2016) found a positive relationship between photography and tourists' levels of happiness. Holidays are often described as a time for relaxing, letting go of work and home stresses, enjoyment and feeling happy (Pearce *et al.*, 2011); they reinforce established social relations, such as with friends and family (Mitas *et al.*, 2012). Relaxation, enjoyment and associated positive emotions during a holiday are also likely to increase levels of life satisfaction, according to the 'broaden and build' theory (Fredrickson, 2001).

Gilovich *et al.* (2015) found that people derive more satisfaction and happiness from experiential purchases than from material purchases, for the following reasons:

(a) Experiential purchases enhance social relations more effectively than material goods.
(b) Experiential purchases form a bigger part of a person's identity.
(c) Experiential purchases are evaluated more on their own terms and evoke fewer social comparisons than material purchases.

The anticipation of a holiday trip may boost happiness (Parrinello, 1993), while positive memories following a trip may also boost happiness, even if the trip was not very pleasurable at the time (Mitchell *et al.*, 1997).

In fact, different types of tourism experiences, leisure and physical activities may boost happiness levels. A number of researchers have noted the healing power of being in nature (for example, in a forest: Konu, 2015; Lee & Kim, 2016), while McCarthy (2016) discussed the well-being benefits of spas. Other researchers have made reference to health-related travel, which may include activities such as spa breaks, seawater treatments and pilgrimages to sacred sites for physical or spiritual healing (Bennett *et al.*, 2004; Hartwell *et al.*, 2018).

For how long does a leisure trip contribute to someone's happiness?

Nawijn and Peeters (2010) argued that personal travel preferences affect tourists' happiness, with a greater number of trips likely to boost happiness even more. However, there have been contradictory findings when it comes to the length of a trip. Kemp *et al.* (2008) found that length of stay does not moderate happiness scores. Etkin and Mogilner (2016) found that whether the variety of activities increases or decreases happiness depends on the perceived duration of the time within which the activities occur; over longer time periods (for example, a day instead of an hour), variety increases happiness. These findings lend support to the saying that 'variety is the spice of life' (albeit not the spice of an hour).

Regarding the question of how happy tourists are during a day of their holiday, Nawijn (2011) found that tourists generally score highly on hedonic levels of affect while on vacation, with positive affect exceeding negative affect. Another study concluded that tourism contributes to short-term well-being only and that tourists' happiness does not increase their long-term well-being (Nawijn *et al.*, 2010). Likewise, De Bloom *et al.* (2010) found that positive vacation effects on the well-being of participants vanished within the first week after returning home. This is because participants resumed their daily routine, causing elevated levels of mental stress.

While on holiday, people may carry with them thoughts that are negative, melancholic and stressful, and this may impact on their travel experience. Nevertheless, sightseeing, social interactions and physical and spiritual activities while on vacation may have a positive influence on their experience and psychology. Anticipation of the trip and the experience itself may become a source of positive feelings, such as 'joy' and 'hope'. However, on return from the trip, they are inevitably faced again with daily challenges and difficulties (Christou & Simillidou, 2020).

Levi *et al.* (2019) investigated the impact associated with engagement in tourism activity on the severity of depressive symptoms in a population suffering from major depressive disorder (MDD) and receiving psychotherapy. Before their vacations, all the participants expressed negative attitudes towards tourism (including unusually high mental stress levels and no desire to disengage from their usual lives). In their post-tourism therapy sessions, their reactions varied, with some reporting that the period of vacation had allowed them to relax, discard negative habits, set future goals and confess to or confront friends and family. In their discussion of these results, the researchers suggested that several aspects of vacations are liable to exacerbate the mental state of MDD patients. Hence, it appears that the effect of tourist activity on depression levels is an individual matter and may be influenced by numerous variables:

- mental stress levels during the vacation;
- the type of vacation (for example, spending time in nature compared to a rushed visit to a crowded city);
- the participant's interests;
- the circumstances of travel;
- the number of travelling partners and their relationship to the participant;
- demographic data (such as socioeconomic status);
- uncontrollable variables (including weather conditions and the availability of facilities and activities suitable for the mental state of the participant).

Hedonistic (short-term) and long-term tourism experiences that contribute to well-being

Despite individual differences, Voigt *et al.* (2011) found that well-being experiences that take place in a spa are likely to be 'hedonistic', whereas more 'eudaimonic' experiences can be gained from spiritual retreats. According to the Global Wellness Institute (2019), more businesses and wellness destinations are taking a radical new approach to 'silence' and 'disconnection'. This means an increased demand for and offering of 'silent spas', silent airports, hotels and resorts with quiet rooms, quiet zones and 'digital kill switches'. Forest-based well-being tourism is founded on the qualities of nature and 'silence' and is a sought-after experience in certain destinations such as Finland (Komppula *et al.*, 2017). Kaaristo (2014) claimed that tourism farmers in Estonia regard silence as one of the most valuable qualities of their rural environments, and other locations appear to have the same appeal:

> On our final morning of the cruise across Doubtful Sound [*New Zealand*], as the incredibly scenic Hall Arm section of Doubtful Sound came into view, our guide urged us to put down our cameras for a moment. To give our full attention to the majesty of nature for 10 minutes of silence. It was a powerful experience. An almost sacred experience that brought tears to my eyes. In an era where most of us too often squander our time clicking through social media and feeling preoccupied with our trivial cares of the day, the wild Doubtful Sound is a powerful and welcome antidote. (Chris, quoted in Explore, 2020).

Spiritual and religious travel is another form of tourism activity that may contribute to well-being and happiness, based on a number of benefits it offers to individuals. For instance, spiritual travel and spiritual activities promote restorative outcomes, increasing the possibility of gaining new perspectives on life, spiritual awakening and cultivation, having a break from a stressful routine, gaining new experiences and engaging in meaningful conversations with interesting people such as elders and spiritual leaders (Christou, 2016; Gill et al., 2019; Gothóni, 2000; Markides, 2017; Olsen & Timothy, 2006). In the case of travelling to spiritual places, participants may consider hotel ratings, extra benefits and price as less important than the actual programme and the essence of the religious trip (Triantafillidou et al., 2010). Thus, spiritual tourism is an important target for those interested in the well-being of people, and further investigation of spiritual tourism may help to understand the strategies employed by people to resolve problems in their everyday experience and reflective assessments (Norman & Pokorny, 2017). For example, Mount Athos, a peninsula in Greece with 20 monasteries, offers opportunities for tranquillity, psychological comfort and spiritual rejuvenation, as visitors report:

> [I]t's perfect for prayer or contemplation, or catching up on your reading. Even if you're not religious, it's a supremely peaceful place to spend a few days ... to my surprise, I found there were some things about Mount Athos that I really missed – the quiet, the solitude, the absence of advertising. As I turned for home, life suddenly looked a lot more garish. I know I'd make a useless monk, but after my brief visit to the Holy Mountain I can entirely understand why some men decide to turn their backs on the modern world. (Cook, 2012)

> [T]hey [the monks] appeared to carry a little old man who couldn't stand on his feet and seemed to be in great discomfort in each move ... I felt great joy being next to him ... He watered and fed me spiritually, and I was receiving with gratitude and joy ... Words were needless. The oxymoron was happening. The close to death old man, to donate live, biological and spiritual, to the 25-year-old ... We exchanged a few sentences, yet had so much depth! (Farasiotis, 2005: 305–306)

The tourism and well-being experiential matrix

Based on the above discussion of the issues linked to tourism and well-being, the following matrix is offered to explain the relationship (see Figure 5.3). The matrix is based on two key dimensions, the first of which (the horizontal axis) represents the following elements:

(i) *Individualistic experiences.* For instance, beauty spa holidays are hedonistic in their nature (Voigt et al., 2011) and address the individual solely.

(ii) *Well-being touristic experiences for the common good.* For instance, volunteer tourism, if conducted in an ethical and appropriate manner, may channel benefits towards others, such as disadvantaged people and regions.

The other dimension (the vertical axis) represents the following elements:

(i) *Well-being experiences that are hedonistic and that may be linked with the body.* For instance, charity (sport) challenge tourism may offer the participant a form of (body) hedonism, while at the same time helping others through the money gathered by the charity event. Charity challenge tourism has been linked with the well-being of participants as well as benefits to certain charity organisations. As Coghlan (2015b) noted, charity challenges sit at the crossroads between charity events, sport and tourism; for example, participants may sign up for a multi-day tour while training to raise funds for a charity. These events contribute to well-being because the participants are active, doing something meaningful and contributing something to those in need.

(ii) *Well-being experiences that are primarily linked with the spirit, psychology and soul of the participant.* Examples include specific wellness experiences, spiritual tourism and nature tourism. Silence and

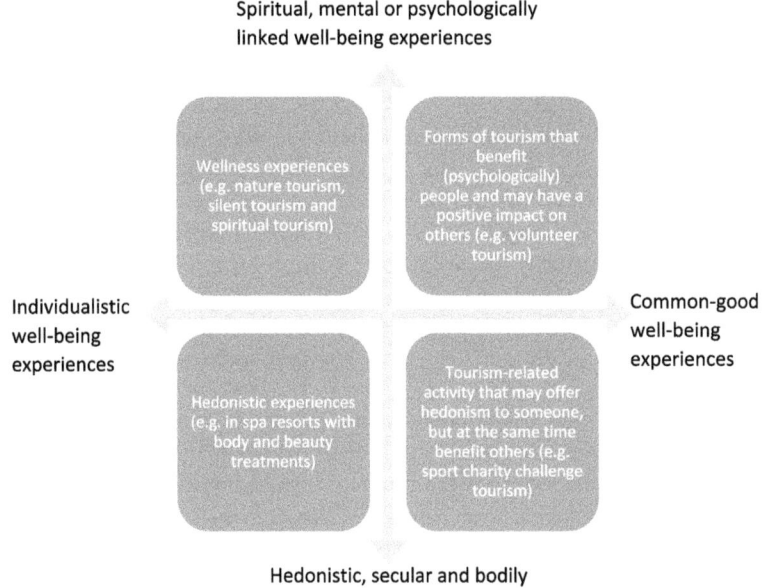

Figure 5.3 The well-being and tourism experiential matrix
Source: Author.

silence tourism are also valued for their beneficial psychological effects on individuals and as contributors to their well-being (Global Wellness Institute, 2019; Kaaristo, 2014; Komppula *et al.*, 2017).

The nexus of tourism and happiness

In order to summarise the main findings derived from the studies discussed above, Figure 5.4 sets out the tourism and happiness nexus. It consists of four interconnected circles which show how tourism may become a source of happiness for participants. There is evidence of a positive connection between happiness and holiday trips (Kemp *et al.*, 2008; Nawijn & Peeters, 2010), and people may derive more happiness from experiential purchases than from material purchases (Gilovich *et al.*, 2015). The tourist is placed in the middle of these interlocked circles: his/her happiness is the result of personal psychological and experiential gains and positive effects on the well-being of others through his/her travel activity. Then, moving from an individualistic perspective to the more social and common sphere of happiness, Smith and Diekmann (2017) discussed the role and nexus of utilitarian philosophy within the context of tourism and quality of life. As they noted, the relevance of this philosophy for quality of life research is self-evident, in that it should promote the maximum benefits for the greatest number of people. The same researchers (Smith & Diekmann, 2017: 9) conceptualised well-being and types of tourism on the following spectrum:

(i) *Short term*: hedonic well-being (for example, sun, sea and sand tourism and hen parties);
(ii) *Medium term*: eudaimonic and hedonic well-being (for example, combining cultural tourism and nightlife, or combining volunteer tourism and beach relaxation);
(iii) *Long term*: eudaimonic well-being, with a list of objectives and existential authenticity (for example, volunteer tourism and spiritual pilgrimage);
(iv) *Permanent and optimum*: utilitarian well-being (for example, sustainable ecotourism and ethical indigenous tourism).

Forms of tourism that may benefit others (instead of only the individual) include sustainable ecotourism, volunteer tourism and social tourism. Changes in assessments of SWB among low-income families were detected in the study by McCabe and Johnson (2013), before and after a social tourism holiday. Despite criticism based on ethical concerns, 'slum tours' were found to generate in visitors a deep feeling of gratitude and life satisfaction for their own situation (Diekmann & Hannam, 2012). Likewise, 'volunteer tourism' can have a strong impact on participants' identity and sense of self (Coghlan, 2015a).

Receiving Happiness 91

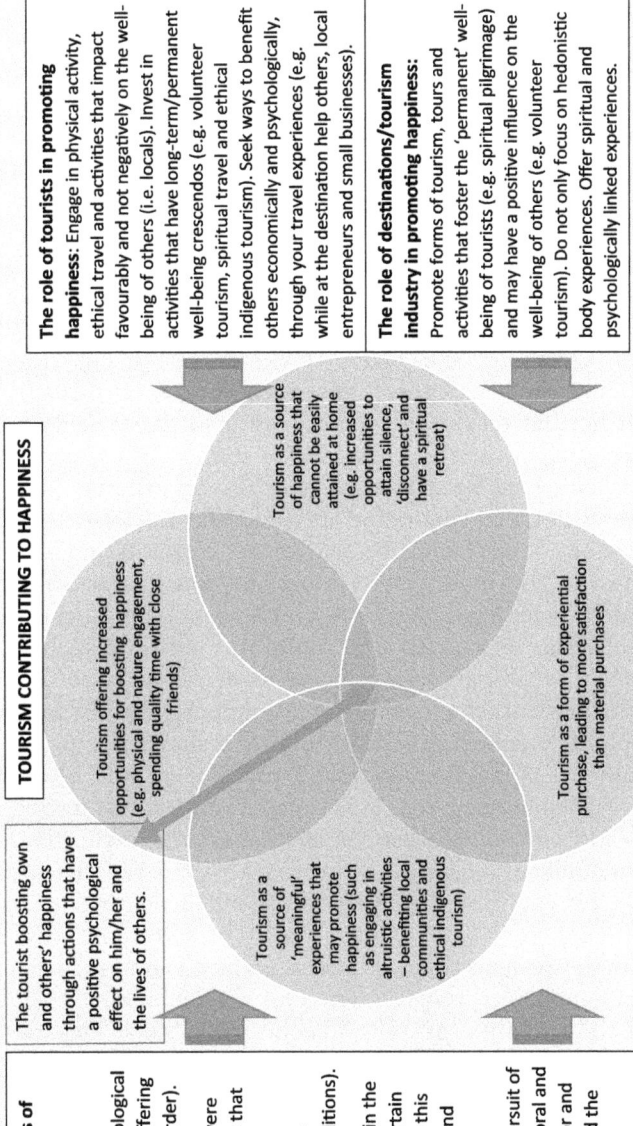

Figure 5.4 The nexus of tourism and happiness
Source: Author.

In terms of Figure 5.4, various dynamics inform and shape the nexus of tourism and happiness. These can be divided into the dynamics of personal, external, spiritual, ethical and travel circumstances. The effects are not necessarily positive. As discussed above, in the study by Levi *et al.* (2019), people who were suffering from MDD expressed negative attitudes towards tourism before their vacations. External and environmental reasons may also disturb the relationship between tourism and happiness; for example, air pollution was found to increase depressive symptoms and reduce hedonic happiness (Zhang *et al.*, 2017).

Tourists, destinations and the tourism/hospitality industry also play a major role in increasing or reducing happiness levels for participants in tourism. For instance, individual tourists may influence their happiness levels by engaging in certain activities such as trekking, given that physical activity is associated with happiness (Khazaee-Pool *et al.*, 2015). Conversely, bad behaviour on the part of some tourists may affect the well-being of others such as service providers. Hu *et al.* (2017) provided evidence that misbehaviour by air passengers was associated with employee role stress, emotional labour and emotional exhaustion.

In an example of how travel companies can impact favourably on happiness levels, Smith and Diekmann (2017) noted the growing number of travel companies offering packages that allow tourists to combine hedonic experiences, such as relaxation, with a meaningful experience, such as cultural and volunteer experiences. These packages are designed to take into account sustainability issues and, perhaps most importantly, the well-being of local people. The same researchers proposed a model of 'integrative well-being tourism experience', located at the intersection of the following interconnected areas: pleasure and hedonism (that is, having fun); rest and relaxation; meaningful experiences (for example, education, self-development and self-fulfilment); and altruistic activities and sustainability (that is, being environmentally friendly and benefiting local communities).

It is clear that there are many difficult questions and dilemmas regarding the 'pursuit' of happiness through travel. This may be referred to as 'the ethical travel and happiness dilemma'. People may seek to increase their happiness levels through travel activity, but their behaviour may challenge the well-being and happiness levels of others, including locals and service providers. People working in the hospitality and tourism industry are often under a lot of pressure and stress, and unruly behaviour by tourists is an extra burden for them. For someone to enjoy a holiday requires other people to make the travel arrangements, prepare the room, organise the tour, walk around the pool to take orders, cook the meal and serve the cocktail.

It is a harsh reality that the tourism industry is often characterised by demanding working conditions, difficult shift patterns and stressful conditions, all of which have a negative impact on the well-being of its employees, such as waiters/waitresses and kitchen personnel (Robinson, 2019). Frontline employees in the hospitality industry may experience tremendous emotional demands during interpersonal interactions, leading to emotional exhaustion (Teoh *et al.*, 2019). Therefore, in our pursuit of our enjoyment and happiness through travel activity, we must consider 'the other' – the other person who works for us, the person who offers hospitality to us, the person who lives there. One should always ask the question, 'How does my pursuit of happiness through travel affect the happiness of others?'

Another question to consider is whether through our pursuit of happiness we are contributing to the well-being of others such as local residents, disadvantaged regions and small local businesses. It should not be forgotten that the hospitality industry is dominated by small and medium-sized enterprises, which are often led by entrepreneurs who face significant challenges in managing their own business and their own well-being (Peters *et al.*, 2019).

However, taking into account the issues of overconsumption and greed discussed earlier in this book, perhaps the most challenging question to answer is this: 'Did I over-do it with travelling this year?' It seems clear that increased travel activity can have a negative impact on the environment, scarce resources and the well-being of others. Morally speaking, the money we chose to spend on our own leisure and luxury holidays could have been used for the common good (for example, by making a donation to someone in need).

6 Over-receiving: Unruly Behaviour, Gluttony and Overconsumption in Tourism

> Everywhere a person blames nature and fate, yet his fate is mostly but the echo of his character and passions, his mistakes and weaknesses.
> Democritus, c.460–c.370 BC

Unruly and 'Carnivalesque' Tourist Behaviour

It is acknowledged that tourism offers increased opportunities for people to escape from their daily routine and stressful lives, to relax, to socialise and to indulge and enjoy themselves.

> Many Americans like to envision vacations as 'time off' – in a sense, the denial of time as one conceives it. One should make love all day, or stay up all night and sleep in the day, denying the 'normal' rituals of the temporal sequence. (Gottlieb, 1982: 170)

This is particularly the case at cultural events, festivals and carnivals (Gotham, 2007; Nurse, 2004; Picard & Robinson, 2006), with carnivals contributing to social cohesion through community celebration (Cuffy, 2017).

The concept of carnival is used beyond its original meaning as a particular cultural festivity (Weichselbaumer, 2012). The concepts of 'carnival' and 'carnivalesque' are linked to the work of Mikhail Bakhtin, a Russian literary theorist who used the concept to denote the festive life of the Middle Ages and the Renaissance (Bakhtin, 1984). The Latin *carne vale* means 'farewell to the flesh', and refers to the last days before Lent, the period of mourning for the passion of Jesus Christ. Despite these origins, carnival as a type of festivity linked with pagan traditions, masquerade, overindulgence, hedonism and sexual promiscuity fails to coincide with the Christian way of living (Evripidou, 2010). During the carnival period in many parts of the world, hedonism and licentiousness are celebrated in dance, masquerading and feasting. The carnivalesque period allows the suspension of traditional rules and a break from the daily work regime (Ryan & Hall, 2001).

Carnival can be regarded as an extension of the Roman festival of Saturnalia, a pagan celebration of the rebirth of the year. The Romans identified Saturn with the Greek Cronus and perceived of him as a god of generation, plenty, wealth, periodic renewal and liberation. Saturnalia ran for seven days from 17 December, and it was a time when rules were turned upside down, with men dressed as women and masters as servants (BBC, 2006). This was a feast period in honour of Saturn, when patricians and slaves would gather in a ritual and liminal space to celebrate a period of peace, abundance and friendship, without private property or slavery. Despite its utopian promise, the festival was nevertheless a temporary state of affairs in which people were also reminded forcibly of the power of the state (Ravenscroft & Gilchrist, 2009). The literature surrounding the carnivalesque and liminality often refers to the subversion of hierarchy and status, and festival spaces are arguably appropriate locations for both the carnivalesque and the liminal (Pielichaty, 2015).

A liminal space is where transformation takes place; it is a space of transition, waiting and not knowing, and a time between 'what was' and 'what is next'. Liminality can be understood as a socially defined area that allows individuals to experience 'freedom' from the mundane existence of everyday life (Shields, 1990) and can include absorption into fantasy worlds (Light, 2009). According to Varley (2011: 86), liminality is useful to express the 'central quality' of leisure experiences. The sociologist Erik Cohen conceptualised tourists as people who travel in the expectation of receiving pleasure from novelty and change, away from the everyday rules of their lives (Cohen, 1974). Indeed, tourism is commonly understood as a period when the normal everyday constraints are suspended, leaving tourists free from the bounds of home and work and in a position to transgress their ordinary 'appropriate' performances (Edensor, 2007). Tourism often appears in both popular and academic literature as shorthand for mass tourism, and denotes a carnivalesque phenomenon that involves gritty vulgarity and a culture of indulgence and playful crowds (Obrador, 2012; Obrador-Pons *et al.*, 2009).

> Touristic experience takes many forms ... but is usually characterised by its extraordinary dimension, temporal limitation, and the absence of responsibility – in other words, by the carnivalesque. (Kennedy, 1998: 175)

The study by Chapman and Light (2017) revealed that employees at a seaside amusement park regularly experienced misbehaviour, such as abusive language and violence, from visitors reflecting the spirit of carnival. Arguably, carnivalesque behaviour may be more easily expressed in settings that give tourists the opportunity to party hard, such as clubs and bars where guests can drink without limitations. Such places are usually in tourist 'enclaves' – specific areas in destinations where tourist activities are planned and where tourists congregate in particular (usually small) geographical areas.

> For years, it's been tourists rather than local residents who have been shaping the image of some of Europe's most beautiful and unique cities. They are being transformed into museums and theme parks and are developing special zones for tourists where locals may work, but certainly don't live. Tourists sit in traditional restaurants devoid of locals as they watch other tourists. They are no longer places where people come together, but where divides seem to deepen. (Der Spiegel, 2018)

The large number of tourists in such enclaves, combined with the unruly behaviour of some, may have negative impacts on the local society. As Dickinson (2018) reported, in 2018 more than 500 people took to the streets of Ibiza to protest against disrespectful and excessive tourism on this Balearic island famed for its hedonistic 24-hour lifestyle. The rally was organised by a local pressure group, and protestors decried the privatisation of beaches, the use of party boats and the rise in crime and noise pollution. The mayors of all five of the Spanish island's municipalities said they wanted an end to intrusion and to the problems that are harming the island's citizens and tourism industry (Osborne, 2018).

Likewise, the coastal town of Ayia Napa in Cyprus developed a reputation for drunken debauchery from the 'Club 18–30' crowd in the 1990s, with groups of young men and women cavorting into the small hours. The mayor of the town, like his counterparts in Magaluf, Ibiza and Hvar, called on the police, tour operators and other visitors to help banish the loutish behaviour of certain tourists (Morris, 2018b).

> We shall not allow a small number of our visitors to turn away the remaining 90 per cent, with their negative actions and behaviours. We would therefore like to make it public that such inappropriate behaviour from certain clientele (low quality youth market) and all their related business partners (tour operators, nightlife event organisers, etc.) are not welcomed in Ayia Napa. This is the one of the 'organised youth' which misbehaves, does not respect the laws of Cyprus, and with their actions creates negative publicity and actually do damage [to] the reputation of the brand name Ayia Napa. (Mayor Karousos, quoted in Morris, 2018b)

Figure 6.1 presents the linkage of enclaved (tourism) areas, liminality and carnivalesque behaviour.

Obviously, not all forms of tourism are associated with carnivalesque disorderly behaviour, nor do all tourists behave badly at specific times of the year, such as the carnival season or during festivities. Unruly behaviour by certain tourists knows no boundaries, and a tourist certainly does not wait for a particular season in order to behave badly.

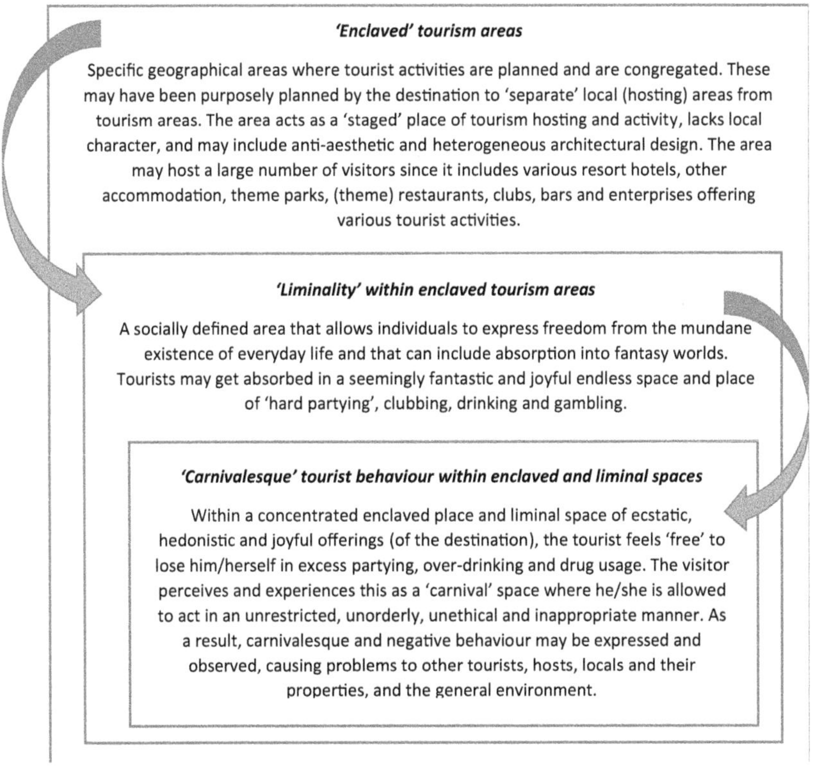

'Enclaved' tourism areas

Specific geographical areas where tourist activities are planned and are congregated. These may have been purposely planned by the destination to 'separate' local (hosting) areas from tourism areas. The area acts as a 'staged' place of tourism hosting and activity, lacks local character, and may include anti-aesthetic and heterogeneous architectural design. The area may host a large number of visitors since it includes various resort hotels, other accommodation, theme parks, (theme) restaurants, clubs, bars and enterprises offering various tourist activities.

'Liminality' within enclaved tourism areas

A socially defined area that allows individuals to express freedom from the mundane existence of everyday life and that can include absorption into fantasy worlds. Tourists may get absorbed in a seemingly fantastic and joyful endless space and place of 'hard partying', clubbing, drinking and gambling.

'Carnivalesque' tourist behaviour within enclaved and liminal spaces

Within a concentrated enclaved place and liminal space of ecstatic, hedonistic and joyful offerings (of the destination), the tourist feels 'free' to lose him/herself in excess partying, over-drinking and drug usage. The visitor perceives and experiences this as a 'carnival' space where he/she is allowed to act in an unrestricted, unorderly, unethical and inappropriate manner. As a result, carnivalesque and negative behaviour may be expressed and observed, causing problems to other tourists, hosts, locals and their properties, and the general environment.

Figure 6.1 The linkage of enclaved tourism areas, liminality and carnivalesque behaviour
Source: Author.

Tourists' violent behavior caused the death of two peacocks at a city zoo in February 2016. The birds are believed to have died from shock after visitors picked them up to pose with them for photographs at Yunnan Zoo in Kumming [China]. Some even plucked out the feathers of the peacocks. (South China Morning Post, 2017)

A tourist's bad and unethical behaviour can take many forms. For instance, two sisters from Arizona were arrested for taking nude selfies at the famous site of Angkor Wat, Cambodia. The pair were inside the Preah Khan temple when they pulled down their pants and photographed their buttocks. In another incident, two Russian tourists shot a raunchy ten-minute video with sexually explicit scenes inside the Giza Necropolis in Egypt (CN Traveler, 2015a). Tourists pushing others to get to the front of the line, urinating in public areas and committing vandalism are just some examples of inappropriate tourist behaviour reported at Shanghai Disneyland Park (McGuire, 2016). In 2013, one of China's four deputy prime ministers, Wang Yang, said that while other countries had welcomed Chinese tourism, the quality of some travellers was not high:

> They speak loudly in public, carve characters on tourist attractions, cross the road when the traffic lights are still red, spit anywhere and [*carry out*] some other uncivilised behaviour. It damages the image of the Chinese people and has a very bad impact. (Wang Yang, in Branigan, 2013)

In Rome in 2015, two young women from California broke away from their tour group to scratch their initials into the ancient amphitheatre of the Coliseum (Scammell, 2015). Another incident was reported by Graham-McLay (2019):

> An English family touring New Zealand – about 12 people, including children – was so unruly that social media posts documenting their bad behavior were picked up by national news outlets, which sent out alerts about the family's location and latest reported antics, including refusing to pick up garbage left on the beach and throwing food on the floor at a café … They dumped chips on someone's beach blanket. They reportedly stole a Christmas tree from a gas station. They were eventually issued deportation notices after the police were called to a disturbance at a Burger King. (Graham-McLay, 2019)

Firm, dubious or unorthodox ways of dealing with (unruly?) tourist behaviour

Certain regulations may be imposed by authorities to control or eliminate what they regard as 'unruly' tourist behaviour. In 2017, Beijing's Temple of Heaven park installed face scanners in the visitor toilets. Those needing toilet paper had to make eye contact with a machine before it spat out a single portion; those who needed more had to wait for nine minutes. The park officials said that this was done to prevent an epidemic of toilet paper theft (Griffiths, 2019). Similarly, a number of Japanese tourist attractions began refusing entry to 'foreigners' in 2019, owing to the unruly behaviour and poor manners of overseas tourists (Dickinson, 2019).

In 2019, the city authorities of Rome imposed a ban at the famous 'Spanish Steps', with the intention of preventing people from sitting on the steps. Their argument was that too many people were sitting there for too long, obstructing the steps for others or eating lunches from nearby fast-food joints. Some people agreed with the ban while others were unhappy about it. Italian art critic Vittorio Sgarbi called the move 'fascist-like' (Pullella, 2019). According to Coffey (2019), the Italian capital has since expanded its list of regulations, including outlawing people going topless in public. Other regulations prohibit the practice of attaching padlocks (with initials, names or love notes) to bridges, eating what the authorities refer to as 'messy foods' around popular attractions (such as the Trevi fountain) and touching one's lips against the spout when drinking from public water fountains.

The Chinese-language booklet originally titled *Common Sense When Travelling in Hokkaido* (Northern Japan) featured numerous examples of bad tourist behaviour. After a Chinese resident made a complaint about the booklet, saying that it implied that visitors do not have good manners and common sense, the Hokkaido Tourism Organisation revised the text. The new version, *Etiquette Guide*, has a more positive spin to it, but it continues to provide points of reference for appropriate behaviour by tourists when visiting this region of Japan (BBC, 2016a): be polite, be quiet, lower your voice in settings such as restaurants and public spaces, act in an orderly way even in long queues, be punctual and do not open the packaging of a product until it has been purchased.

The Role of Overconsumption and 'Gluttony' in Tourism

Everything in excess is opposed to nature. (Hippocrates, *c*.460–*c*.370 BC)

Aristotle encouraged moderation in all things and held that the extremes are degraded and immoral. In this sense, courage could be regarded as a moderate virtue located somewhere between cowardice and recklessness. Moderation seems to be in sharp antithesis to overconsumption. The global setting is being progressively impacted by the dominance of mass tourism (Harrison & Sharpley, 2017; Hernández *et al*., 2016; Lai & Hitchcock, 2017) and increased tourist demands for food, services and tourism experiences.

The hospitality and tourism industry has been strongly criticised for the amount of food waste it generates (Filimonau & De Coteau, 2019), and the sector's demand for food has significant impacts, especially for small destinations such as islands (Pratt, 2013). Negative impacts affect the environment, the ability of host communities to provide necessities and obtain food, and the emotional states of local people (Jordan *et al*., 2019; MacNeil & Wozniak, 2018) (see Photo 6.1).

Alongside the continuing increase in tourist numbers and mass tourism phenomena, there are other reasons for overconsumption in tourism, such as personal motivational factors, the continuous search for hedonistic experiences and pleasure for the bodily senses, undisciplined consumption and the promotional efforts of destinations. Young *et al*. (2014) noted several different forms of undisciplined and excessive tourist consumption causing negative individual and societal impacts.

Overconsumption in tourism: A motivational and hedonistic perspective

Tourism motivation is a fascinating yet perplexing and complex phenomenon (see Terzidou *et al*., 2018) which can partly explain overconsumption in tourism. It is acknowledged that motivation for travel functions as the driving force behind the behaviour of tourists, while the travel motives of individuals are influenced by a range of personal and

Photo 6.1 Tropical beach, Thailand. Destinations may unintentionally incite mass and overtourism phenomena in particular places by including beautiful settings in their marketing campaigns, like this picture taken on Thailand's tropical coast
Source: Author.

external factors (Crompton, 1979; Farmaki *et al.*, 2019). Despite this complexity, push/pull motivation emerges as a theory that dominates much of the literature on tourist motivation. Push factors are acknowledged as internal forces, such as the emotional needs that shape an individual's decision making and reasons for travel; for instance, push factors include fun, education, escape, novelty, family togetherness, excitement, nostalgia seeking, self-development, relaxation and luxury (Christou *et al.*, 2018b; Jamrozy & Uysal, 1994; Jang & Wu, 2006; Yoon & Uysal, 2005). Pull factors are attributes that destinations use in marketing to attract visitors (Dann, 1981; Goossens, 2000). A recent research stream has acknowledged the role of hedonism seeking in ethical and volunteer tourism (Coghlan, 2015a; Malone *et al.*, 2014).

The list of tourist motivations seems boundless, as it embraces everything from the need for spiritual awakening and fulfilment (Terzidou *et al.*, 2018) and mythical pursuits (Christou & Farmaki, 2019; Laing & Crouch, 2011) to even the desire to consume the 'dark' components of tourism (Podoshen, 2013). Nevertheless, tourist motivation may well be underpinned by an urge to satisfy the bodily senses (Christou, 2020b). For instance, people may seek gastronomic travel experiences, driven to a specific destination by a longing to savour specific delicacies (Ellis *et al.*, 2018; Kim *et al.*, 2019) (see Photo 6.2). Tourism has long been heralded as a pleasure-seeking motivator and/or a promoter of self-indulgent and hedonistic experiences (Bigne *et al.*, 2005; Goossens, 2000), and eating is a form

Photo 6.2 Food market, Stockholm, Sweden. Tourists' gastronomic experiences range from savouring local delicacies to visiting particular destinations for a specific indulgent gastronomic experience, as with wine and cheese tours in wine-producing regions. In this photo, tourists visit a famous traditional food market in the city of Stockholm
Source: Author.

of tourist activity that gratifies all five senses and fulfils the experiential part of the tourist experience by offering pleasure (Mak *et al.*, 2013, 2016).

A study by Koc (2016) investigated the extent of the illusion of control as a form of cognitive bias that tourists may have when making decisions about food consumption while on holiday. The study concluded that all-inclusive holidays with free open buffets cause tourists to enjoy their food and beverages more, and may often cause them to engage in gluttonous behaviour. At the same time, tourists may be led to overconsumption by excessive offerings and by mimicking the behaviour of others:

> Eating and drinking on an all-inclusive holiday is a bit like a competition. You see others doing it and you feel like you should not be left behind.
> (Koc, 2013: 830)

The role of 'gluttony' in overconsumption

> If you are ruled by the mind you are king, if by the body, you are a slave.
> (Catherine of Alexandria, *c*.287–*c*.305 CE)

The word 'gluttony' derives from the Latin *glutto*, related to *gluttire* (to swallow), *gluttus* (greedy) and *gula* (throat). Although gluttony can manifest itself in other forms, it basically implies excessive eating and drinking and excessive attention to food and drink. It implies a predilection for food that has enjoyable flavours or the consumption of food exclusively for pleasure. Gluttony has been linked to overconsumption within the realms of food, alcohol and the spending of money (Unger & Scherer, 2010; Veselka *et al.*, 2014).

> A video of tourists using their plates to shovel shrimp at a hotel buffet in Chiang Mai, Thailand, left internet users around the world reacting with horror. Subsequent photos showed most of the food was left uneaten and wasted. The tourists were seen jostling with each other noisily to get at the food, and walking away with several overflowing plates each. (South China Morning Post, 2017)

Gluttony may also imply overindulgence in (wealth) elements. '[It] is the overindulgence or overconsumption of anything – not just food – to the point of waste' (Dossey, 2010: 3). This is the main reason why for centuries many cultures have frowned upon gluttony. Gluttony has been stigmatised mainly for its egocentric tendency towards excessive care/pleasure of oneself, often at the expense of consideration of others. It has also been associated with waste and environmental vice (Cafaro, 2005; Edwards & Mercer, 2007) and with feebleness of spirit, weakness of will, vulnerability to passion, hedonistic-seeking behaviour and over-attention to pleasing one's mouth and stomach. Gluttony was once regarded in Europe as a deadly sin – 'Let him be damned, like the glutton!' (Shakespeare, *2 Henry IV* 1.2.35–36) – while in medieval Japan it was conceived as the karmic consequence of a moral failing in a Buddhist context (Stunkard *et al.*, 1998; Veselka *et al.*, 2014).

Gluttony has been referred to as a vice (Miller, 1997) and related to overindulgence, pleasure seeking and bodily motivated behaviour towards food, alcohol, sex and drugs (Hull, 2011; Vrabel *et al.*, 2019). However, gluttonous behaviour is not necessarily restricted to humans, as shown by the case of blood-sucking arthropods (for details, please refer to Kaufman, 2007).Gluttony has been linked to notions of 'excess' (Lindner, 2016) and overconsumption: 'Their [referring to a specific group of people] thefts, violence and gluttony exceed conventional habits of consumption ... who have "no shame nor respect"' (Sykes, 1999: 157). The term is often used along with other related terms, such as sloth, although sloth implies under-exertion and lack of motivation, whereas gluttony implies overconsumption, over-nutrition and overindulgence in pleasures (Miller, 1997; Unger & Scherer, 2010; Veselka *et al.*, 2014); for example, 'when drinking alcohol, I almost always drink to the point of intoxication' (Vrabel *et al.*, 2019: 86).

Gluttony, like sloth, is associated with obesity (Prentice & Jebb, 1995). According to Irvine (2010), gluttony could be regarded as the basis of the obesity epidemic. Approximately half a century after the start of the gastronomic revolution, Unger and Scherer (2010) noted that two-thirds of Americans were overweight or obese. Gluttony has been associated – and often confused – with the terms 'obesity' and 'greed'. In fact, the three terms differ in their nature; gluttony is over-attention to and overconsumption of food, drinks, wealth items and status symbols, whereas obesity, in the strict sense, is a medical condition (a person may be considered as 'obese' if his or her body mass index exceeds 30 kg/m^2). Furthermore, greed refers specifically to an insatiable longing for material gain, perhaps for food, but maybe for money and power. Gluttony has been characterised by overindulgence in extravagant spending and drugs (Miller, 1997) as well as sex. There are many cases within the tourism context of travellers seeking such self-indulgent and/or hedonistic experiences. Examples include those who are primarily or indirectly motivated to travel in order to consume commercial cannabis (Wen *et al.*, 2018), indulge themselves at spas (Dimitrovski & Todorović, 2015; Smith & Diekmann, 2017) or experience commercial sex (Ying & Wen, 2019). Visitors may behave differently in an environment that is dissimilar to their usual one, and they may engage in overindulgence and excessive consumption of food, sex or other bodily activities.

Obstacles to overconsumption

It is widely acknowledged that income and price elasticities have an impact on tourist consumption (Fleischer & Rivlin, 2008). Furthermore, personal, ethical, religious and cultural norms may present obstacles to self-indulgence and excessive (food) consumption. An example is depriving oneself of a bodily pleasure, such as food, in order to offer it to another person in need (Christou, 2018). Other examples are tourists who are taking care of their physical appearance, following a diet (Small, 2016) or observing a period of fasting (Mujtaba, 2016).

From a philosophical and spiritual perspective, virtues such as prudence, fortitude (Prose, 2003) and abstinence have been advised as a means to counter gluttony. In addition, individual barriers to certain consumption activities may include feelings of embarrassment and shame (see Poria *et al.*, 2019) and emotional feelings of discomfort due to the stigma of being 'fat' (Small & Harris, 2012).

Overconsumption: A destination offering (pull) perspective

Within a consumerist society, organisations attempt to influence and arguably support overconsumption, often through their promotional tactics. Aydinoğlu and Krishna (2010) proposed the notion of 'guiltless gluttony', arguing that food vendors adopt certain types of descriptions that may lead to the consumption of larger quantities without consumers being aware of it. A simple search for descriptions on the official websites of destinations, large tour operators and (mostly) upscale resorts reveals that these tend to set the scene for luxurious and extravagant gastronomic experiences for guests. In particular, food and gastronomic experiences are used by destinations as an important marketing tool (Koc, 2013; Okomus *et al.*, 2007).

(i) The senses: Stimulation and (body-related) temptation

The vocabulary used on websites, especially for luxurious five-star hotels, cruise ships and restaurant venues, betrays an attempt by organisations to stimulate the human senses and to encourage visitors with tempting descriptions. Examples of terms they may use to describe food or experiences include: 'delicious', 'mouth-watering', 'succulent', 'stimulate', 'decadent', 'feast', 'culinary journey', 'self-indulgent', 'incomparable, exceptional/unforgettable [meal experience]' and 'wanderlust'.

> The ... package offers the best all-inclusive value at Savoy. It includes delicious buffet-style breakfast ... Guest rooms are stocked with beverages once per stay ... The package also grants daily spa & gym access and includes a number of sports, activities and entertainment offerings. And younger guests can enjoy kid's clubs, free ice cream and complementary ice skating sessions. (Savoy-Sharm El Sheikh, 2019)

> Our luxury beach resorts, set along the most gorgeous tropical settings and exquisite beaches in Saint Lucia, Jamaica, Antigua, The Bahamas, Grenada and Barbados, feature unlimited gourmet dining, unique bars serving premium liquors and wines, and every land and water sport ... (Sandals, 2019)

> The best nights out start with an incredible meal and dining experience. At Wonderland Imaginative Cuisine, master chefs walk the line between fantasy and reality, serving up creative cocktails and dishes designed to whisk you away down a rabbit hole of wonder. (Symphony of the Seas, cruise ship, Royal Caribbean, 2019)

In particular, upscale hotels (such as luxury establishments) emphasise their provision of experiences that include a largely bodily component. Descriptions on their official websites evoke luxurious, lavish and/or hedonistic experiences that feed or comfort the body and its senses, all provided with the least possible physical effort on behalf of the guest. The following excerpt also fits within the rationale of 'all-inclusive' and 'unrestricted' access (as explained below), since it promotes the idea of 'infantilisation' (tourists being treated as children; see Dann, 1995):

> There's nothing worse than getting sun cream on your favourite lenses when you're relaxing poolside, we hear you ... you'll find dedicated sunglass butlers who will pop over and give your favourite sunglasses a polish, so you don't have to. (Jumeirah Hotel, Dubai)

(ii) The unrestricted offering

Overconsumption and overindulgence may be partly explained through exposure to a variety of menu offerings (Poria *et al.*, 2019) and the often unlimited access to food/activity provisions in 'all-inclusive' options (Farmaki *et al.*, 2017). All-inclusive packages and experiences are offered worldwide, especially in destinations that rely heavily on tourism, such as the Canary Islands in Spain (Zoghbi-Manrique *et al.*, 2013) and Cape Verde (López-Guzmán *et al.*, 2016). An all-inclusive package is one in which visitors enjoy almost unrestricted consumption of food and beverages within the hotel, subject to certain time limits. The visitor pays upfront, ensuring flexibility and often unconstrained consumption of meals, drinks and (hotel) entertainment, from basic to extravagant offerings. In such cases, guests may have all their needs met by the hotel, including social activities (Zoghbi-Manrique *et al.*, 2013).

In these kinds of circumstances, people may consume more, in terms of both quantity and variety, while on vacation (Koc, 2013). The offering of unlimited choices and unrestricted food/beverage consumption is particularly prominent on the websites of buffet restaurants, hotels or cruise ships offering all-inclusive options, with references to the enormous quantities of food and large eating spaces available: 'unlimited', 'huge [buffet]', 'generous [portions]', 'endless', 'an array of [dishes]' and 'buckets'. Similarly, mega malls at specific destinations tempt visitations with descriptions of extravagant shopping experiences and a plethora of shopping choices.

Tourism and overconsumption: A conceptual association

Figure 6.2 sets out clearly the interrelated nature of tourism and overconsumption. In the tourism experiential context, people may become over-dependent on travel, tourism and gastronomic experiences that involve excessive attention to satisfying the bodily senses, pursuing materialism and consuming food. Examples include excessive consumption of shopping and gambling (Young *et al.*, 2014), as well as of food. As explained in the previous section of this chapter, excessive consumption may be the result of various individual factors. For instance, researchers have reported factors related to genetics, insulin resistance and abdominal obesity, as well as hypothalamic abnormalities (Unger & Scherer, 2010). There are also psychological factors that may lead someone to overconsumption of food (Botsari, 2014).

Over-attention to stimulating and comforting the bodily senses and the desire for hedonistic experiences may well drive the overconsumption of travel and tourism experiences. Of course, not all hedonistic actions can be blamed for overconsumption, and tourists may pursue hedonism while being involved in ethical and volunteer actions (Coghlan, 2015a; Malone *et al.*, 2014). At the other end of the spectrum (on the offering side), tourism providers such as travel organisations, hotels, restaurants and shopping malls attempt through their offerings and their marketing initiatives to stimulate demand for their services/products. These organisations may provide an unrestricted and excessive supply of services,

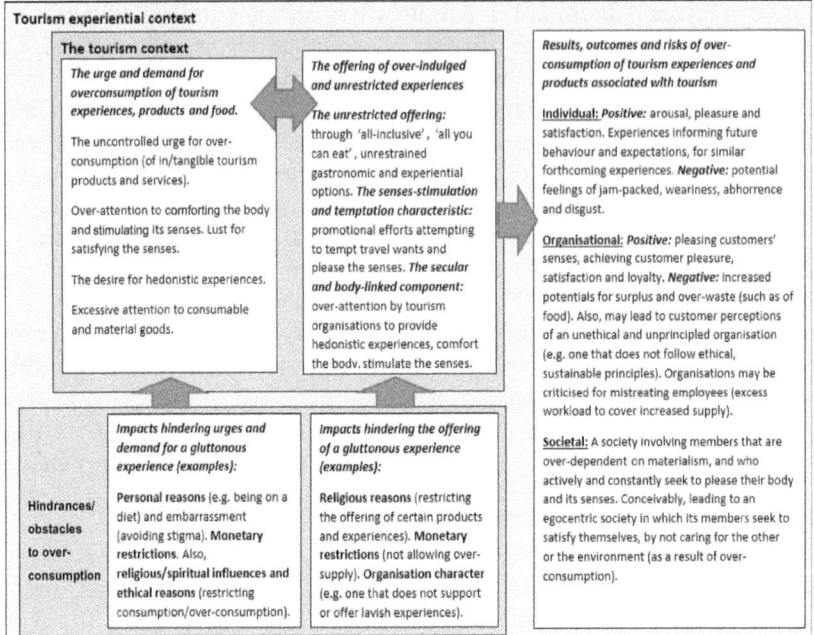

Figure 6.2 Tourism and overconsumption
Source: Author.

products and foods in their attempt to satisfy their customers. Despite the continuing rhetoric of the efficacy of various destination and venue marketing methods, campaigns and branding (March, 1994; Pritchard & Morgan, 2001; Tussyadiah & Park, 2018), such promotional methods have been, and in all likelihood will continue to be, used by the tourist industry.

In this interrelated and dynamic process of motivation, urges, offerings and stimulating factors, there seem to be certain impacts that may hinder both demand for and offerings of gluttonous experiences. Examples include avoiding overindulgence in food in an attempt by an individual to avoid social stigmas and embarrassment, or people taking care of their physical appearance (Small, 2016). Other impacts include religious and spiritual influences which may prevent consumption of certain experiences (such as commercial sex) or overconsumption. A fasting period (Mujtaba, 2016) is an example of a factor that may hinder the offering and consumption of food, at least for a time. Likewise, Elder Porphyrios, an Orthodox spiritual leader urge people to fight the passion as he called it of gluttony (Porphyrios, 2005) and to avoid food hedonism.

Overconsumption and overindulgence do not come without their share of negative impacts. Tourism has been criticised for creating substantial negative environmental impacts, such as altering the sense of a place (Christou *et al.*, 2019c) while reducing the ability of locals to provide necessities and obtain sufficient food (MacNeil & Wozniak, 2018). As a result, destinations and tourism organisations that promote overconsumption, such as all-inclusive packages, may be viewed as unsustainable and unethical since they may put excessive and unnecessary pressure on the environment and the local community. Negative impacts are distributed among individuals, organisations and the society as a whole. Nevertheless, pleasing guests by offering preferred tourism experiences may lead to satisfaction, attachment and/or loyalty (Beattie & Schneider, 2018; Bigne *et al.*, 2005; Radojevic *et al.*, 2018).

In their attempt to meet the extravagant and excessive demands made by some visitors, organisations may channel the pressure to produce/offer more onto their employees. On all-inclusive vacations, tourists consume more in terms of the variety and quantity of food (Koc, 2013), and this creates more occasions when they can witness management's maltreatment of employees or even of themselves (Zoghbi-Manrique *et al.*, 2013). Linked to this is the tourism industry's increased demand for (and possible waste of) food (Filimonau & De Coteau, 2019; Pratt, 2013). In addition, Young *et al.* (2014) emphasised the negative individual and social consequences of undisciplined consumption. Overconsumption can lead to feelings of saturation and possibly disgust. The emotion of 'disgust' can be evidenced by visitors during their trips or experiences (Christou *et al.*, 2018b; Lin *et al.*, 2014), and is regarded by De Jong and Varley (2017: 218) as 'the Other of desire'.

Part C

The Philosophy of Giving and Receiving (Tourism) Places

7 Giving and Receiving Places: The Significance and Spirit of Places, and Tourism Development

> Place is the greatest thing, as it contains all things.
> Thales of Miletus, *c.*624–*c.*546 BC

'Topophilia' as Affection for a Place

Human interaction with the surrounding space (physical and cultural) is at the core of tourism, particularly for specific nature-based and cultural forms of tourism such as ecotourism, geotourism, rural tourism and cultural-heritage tourism. In these cases, the landscape provides natural and cultural assets for tourism development (Stoffelen & Vanneste, 2015) and can shape visitor experiences and perceptions; it may be developed to suit visitor needs, and managed and marketed to increase visitor engagement and enjoyment (Carneiro *et al.*, 2015; Huang, 2013).

Deeper understanding of the link between landscape and tourism is fundamental for the tourism sector in general (Smith & Ram, 2017). Places and their attributes can evoke strong emotions, whether positive or negative, among inhabitants and visitors. Emotions are also provoked by being in a place that is relatively unfamiliar (Urry, 2016). According to Meyer (2003), particular places, such as a national park that is unique and rich in history, may have a spiritual element attached to them and may evoke 'awe' in visitors.

'Place' is more than the geographic location of visitor activity. It is an amalgam of destination qualities such as landscape and architecture, heritage, history, social structures and relationships (Smith, 2015). A place may bring its visitors 'astonishment', 'excitement', 'awe' and 'joy' (Christou, 2019b). The Chinese geographer Yi-Fu Tuan examined both spaces and places and argued that a place is security, freedom and a centre of felt value where biological needs are satisfied (Tuan, 1979, 1980).

People may feel a strong attachment to a particular place such as the rural English Lake District (Jepson & Sharpley, 2015), or they may be emotionally bonded with a particular place, resulting in 'topophilia'. According to Knapton (2017), when people are shown pictures of places that are important to them, there is a far greater boost of activity in the amygdala – a key area in the brain for processing emotion – than when they are shown pictures of important objects. The actual term 'topophilia' was coined by the poet W.H. Auden in 1948, who described it as 'the sense of belonging' people experience when returning to an important place from their past.

> The word 'topophilia' is a neologism, useful in that it can be defined broadly to include all of the human being's affective ties with the material environment. These differ greatly in intensity, subtlety, and mode of expression. The response to environment may be primarily aesthetic … More permanent and less easy to express are feelings that one has toward a place because it is home, the locus of memories, and the means of gaining a livelihood. (Tuan, 1990: 93)

According to Meyer (2003), each place that has meant something to a person touches that person individually. For some people, even the name of a particular place may conjure up sounds, smells and the 'sense' of that place.

'Sense of Place' and *Genius Loci*

> The term 'Sense of Place' is often used to describe your feelings for a place, and the elements that make that place special to you – it may be memories of past visits, views, sounds, people, tastes, even the smell of the place! (Forest of Bowland, 2015)

'Sense of place' has been regarded as a complex and multidimensional concept (Jorgensen & Stedman, 2006; Mullendore *et al.*, 2015). A number of researchers from different disciplines, including the tourism field, have attempted to shed light on what the concept comprises (Kerstetter & Bricker, 2009; Tan *et al.*, 2018). There are multiple discourses about what sense of place is (Chapin & Knapp, 2015). For instance, from a human geography perspective it is regarded as a combination of social constructions interacting with physical settings (Campelo *et al.*, 2014; Liu & Cheung, 2016). Basically, sense of place refers to a characteristic that certain places have and some do not, and to people's perceptions of that characteristic.

According to Hausmann *et al.* (2016), sense of place includes people's perceptions and interpretations of the environment. It also refers to the characteristics that make a place unique and that foster a sense of human attachment as well as belonging. Places that have a strong sense of place possess a strong identity that is felt deeply by people, whether locals or visitors (Bloom, 1990). It should be stressed that a sense of place may carry a positive vibe and feeling or a negative one, such as 'fear' (Tuan,

1980). In 2005, as part of its work on sustainable tourism, the Forest of Bowland Area of Outstanding Natural Beauty in England started a project using the concept of a 'Sense of Place'. They published a useful toolkit (Forest of Bowland, 2019) with the aims of promoting the special qualities of the forest, providing consistent messages for all partners to use, increasing awareness of what the area has to offer to visitors, developing a greater understanding of the geographical area and creating a loyalty to this area in residents and visitors.

Sense of place reflects the ancient concept of *genius loci* (Lecompte *et al.*, 2017), although the latter may carry a stronger spiritual element. In classical Roman religion, the *genius loci* was the protective spirit of a particular site. As Meyer (2003) suggested, in the same way as people have different personalities, places may also have personalities: they are unique, steeped in tradition, rich in history and carry a 'spirit'. Through the accumulation of the experiences of all those generations who have lived there, a place gains a soul and a spirit (Neri, 2006).

In the past, *genius loci* carried a rather negative connotation, with pagan roots and its depiction in Roman religious iconography as a fearsome snake on an altar. The snake acted as a protective symbol, shielding a specific place (Gayley, 1893). More recently, Christou *et al.* (2019c) found that the following specific elements and rudiments were linked to a site's *genius loci*:

(a) the physical, inner and sensory environment: that is, the physical static environment (old buildings), the natural environment (endemic flora) and objects and items in the static environment (wooden seats and icons);
(b) the sensory environment (such as ethereal essences and candles): the creation of an experience that addresses all human senses;
(c) the human interrelationship factor: the social element (for instance, allowing and promoting social interactions with spiritual people);
(d) the aura, spirit and soul of the place: that is, the psycho-spiritual offering of the place.

This last element is the deepest and most challenging aspect to maintain and promote, as personal factors (the interpretations and perceptions assigned to the site by each visitor) play an important role in shaping the aura. Even so, visitors may be encouraged to experience the deepest and most genuine aspects of a site (for example, by taking part in an authentic event or ceremony).

The Alteration and Obliteration of a Place's *Genius Loci* and the Concept of 'Placelessness'

According to Street (2018), every week seems to bring new confirmation that we have officially entered the era of overtourism. Of course, this statement was made prior to the COVID-19 'coronavirus' outbreak which resulted in tourism running dry in famous destinations around the globe (Legorano, 2020). Nevertheless, prior to the pandemic, a noteworthy number of international places experienced penetrating growth in tourist numbers, often accompanied by intense tourism-related development (Christou *et al.*, 2019c; Gursoy *et al.*, 2009; Ioannides, 2008; Kozak & Rimmington, 2000; Lange, 2015; Luong, 2015; Street, 2018; White, 2019). In coastal areas of Zanzibar, rapid expansion of tourist-related infrastructure resulted in dramatic coastal environmental degradation (Lange, 2015), and similar changes have been reported in New Zealand:

> [F]or many New Zealanders, and certainly many residents, Queenstown has got too big, too busy. Used to be great, they grumble. Used to be beautiful. Not like that now. Bloody tourists, they mouth. (White, 2019)

> Wilderness is easily harmed by the very industry [*tourism in New Zealand*] that lauds those qualities. I don't know how we've ended up in a place where saying words like 'economic growth' and 'jobs' and so on carries the moral high ground. (Finlayson, Federated Mountain Clubs, in White, 2019)

Islands in Thailand were reported by Vater (in Haines, 2016) to be under increasing threat from tourism-related overcrowding and the degradation of the environment and landscape:

> Some of the country's most stunning locations, such as the well-known Phi islands, have long succumbed to serious degradation due to too much commerce and resort construction.

Fjaðrárgljúfur Canyon in Iceland became a popular travel destination after it was featured in the television series *Game of Thrones* and in a music video. The sudden influx of people damaged the site, prompting a ban on tourists (Diamond & Olito, 2019). Plog (2001) noted that some destinations follow a pattern of uncontrolled tourism development that may alter their physical settings. In this connection, Der Spiegel (2018) reported the following about certain Mediterranean resorts:

> There were times when the hotels lining the beaches in Benidorm, in Arenal on Mallorca and along the Adriatic Sea in Italy, were symbols of the ugliness of modern mass tourism ... Benidorm and Arenal are cities that were created so that Europeans would have a place to lie on the beach in summer. They are artificial resorts and not very nice.

Impacts on places do not only affect the environment and the quality of life of locals but may also shape the experience of visitors. In a study by Pietilä and Fagerholm (2016), visitors at a national park in Finland were asked to indicate specific locations where they felt that the effects of tourism disturbed the quality of their experience. The visitors reported that crowding and erosion disturbed their experience along highly visited sections of the trail. Across America, national parks and public areas are facing a crisis of popularity as a result of domestic and international tourism, technology and 'successful' marketing. The result is a negative effect on the physical setting, the sense of place and visitors' experiences:

> Backcountry trails are clogging up, mountain roads are thickening with traffic, picturesque vistas are morphing into selfie-taking scrums. And in the process, what is most loved about them risks being lost. (Simmonds et al., 2018)

Of course, it should be stressed that tourism alone is not to blame for every change to a place and its sites. For example, in 1801 Lord Elgin stripped the ancient sacred site of the Parthenon (in Athens) of many of its sculptures and took them to England (Sánchez, 2017). However, the typically rapid and uncontrolled nature of tourism development is particularly associated with the 'paramorphosis' (changing beyond recognition) of places. Paramorphosis is the transformation of a physical setting as a result of extensive tourism development and associated activity (Christou, 2019b), and it can be understood in the context of Butler's widely used TALC (Tourism Area Life Cycle) model of destination development (Butler, 1980), which suggests that destinations go through various stages, starting from birth and then growing in the form of a wave or a circle (Butler, 2009).

A noxious combination of rapid and uncontrolled tourism development may lead to several negative impacts. Attractions may lose their iconicity as a result of environmental deterioration (Weaver & Lawton, 2007; Weidenfeld, 2010). Entertainment facilities, pseudo-native stores and high-rise hotels begin to dominate the original architecture. The place gradually takes on a more 'touristy' appearance, its characteristics are eroded and it loses 'its distinctive character along the way' (Plog, 2001: 19).

There are many examples of places around the world that have lost their identity mainly as a result of tourism:

> It's doubtful, however, whether the Boqueria food market on La Rambla [*Barcelona*] can be saved … the stallholders are one by one caving in to the force majeure of tourism, with fresh fish, meat and vegetables giving way to juice bars and assorted takeaways. The very reason for visiting la Boqueria – even as a tourist – will soon cease to exist. (Burgen, 2018)

> The market [*Bloemenmarkt in Amsterdam*] dates back to a time when the shops would receive their flower shipments directly on the canal and sell them right on the barges, but now most of the shops have been turned into cheap souvenir stalls. (Diamond & Olito, 2019)

Rapid tourism development transformed the once-small village of Ayia Napa in Cyprus into a well-known destination catering mainly for party seekers and 3S (sun, sea and sand) visitors (Ioannides & Holcomb, 2003; Saveriades, 2000; Sonmez *et al.*, 2013). In a study by Christou *et al.* (2019c), respondents made the following points:

> [A]ll these buildings block the view ... all you can see now are huge blocks of cement ... It's not the same [*referring to the setting*]. It has changed, and I personally don't like it. They [*the authorities*] shouldn't give permission to build them. I prefer how it was in the past ...

> They [*the authorities*] should demolish everything and start building again! Unfortunately, now they cannot do much ... it will take them years ... But they can start by investing more money in cleaning the place. I believe that they should not allow commercial signs, or at least make them look more traditional ... They must expand what they've done in the monastery [*preservation*] to the surrounding area ... they should construct more picturesque roads, and relocate elsewhere the clubs and bars surrounding the monastery. (Christou *et al.*, 2019c: 23–24)

In Photo 7.1 a resort hotel in Thailand is built to reflect the character of the place.

Photo 7.1 Phuket Island, Thailand. Hotels should reflect the character of the place through their architecture, design, building materials and gardens. In this way, the appeal and sense of a place is shielded from excessive change. Developers and stakeholders must take into consideration the fragility of the surrounding ecosystem and the local community, allowing locals to 'take in' and sense the environment without privatising the surrounding areas, such as the beach
Source: Author.

The concept of 'placelessness'

Placelessness is the loss of a place's essence (Terkenli, 2002). Relph (1976) argued that lack of consideration of the significance of place in urban development has resulted in the casual eradication of distinctive places and the making of standardised landscapes. Coastlines and natural areas are turned into places that can be more 'tourist-friendly' and recognisable to tourists in search of resorts, hotel chains, branded restaurants and clubs.

> The first time I came back [*to Phu Quoc island in Vietnam*], I marvelled at its coastlines and just how beautiful, raw and natural it was … Now bulldozers are cutting up this raw beauty to make it into something more recognisable to tourists – familiar global hotel chains and resorts. (Trinh in Luong, 2015)

> … the city [*Barcelona*] is rapidly losing its identity and becoming like everywhere else. A new word has been coined to describe this apparently unstoppable process: *parquetematización* – the act of becoming a theme park. Barcelona has become an imitation of itself. (Burgen, 2018)

Likewise, Bates (2012) reported that after tourism development, parts of Cyprus's landscape are now like any other Mediterranean destination, with tourist-friendly shopping malls and international restaurant chains. This is in sharp antithesis to the country's hinterland, which has kept its identity, character and soul.

> [T]he south [*of the island*] had seemed familiar, with only occasional signs bearing the Greek alphabet giving a clue that we were not in any other Mediterranean resort of indistinct origin … I turned my back on the delights of sea and sand and drove into the foothills of the spectacular Troodos Mountains. There among the traditional villages I found Byzantine churches packed full of golden icons and paintings, and, rather touchingly, still very much part of village life. My favourite was in Omodos, among whose tiny cobbled streets I found the Holy Cross monastery. (Bates, 2012)

Some East Asian cities that have created urban landscapes characterised by skyscrapers and large cutting-edge shopping malls with chain restaurants and brand shops, despite attracting an increasing number of visitors, have been criticised for producing environments that lack uniqueness and local flavour (Morley, 2009; Shim & Santos, 2014; Wu, 2003). Shim and Santos (2014) investigated place and placelessness in the phenomenological experience of shopping malls in Seoul:

> Largely, study participants described their experiences by referencing several features that are thought to contribute to placelessness. Specifically, many compared their experiences in Seoul's malls to one in their home country and described the similarities between them as largely centered around such topics as homogenized interiors, multi-national brands,

cosmopolitan settings, similar foods, or American culture. (Shim & Santos, 2014: 109)

> This is not exactly Korean culture and this is not exactly Western culture either. This is kind of anonymous. I don't think there is anything genuinely authentic here. Every time I come to this place there is something new and it keeps on changing. (Participant 23, quoted in Shim & Santos, 2014: 111)

Despite these concerns about excessive change, it should be stressed that it would be over-simple to regard culture as something that should remain fixed. Tourism may bring change to the culture of a place, and the change may be in the interests of at least some of the local population:

> [T]he traditional identity of the region has faced new challenges from debates regarding tourism development, globalization, and commercialization both within and beyond the community. The demand for tourism development and the need to develop into a global city urge Kangnung toward a new identity influenced less by local tradition and more by global culture. These debates can empower marginalized groups, such as women and youth, who value innovation and the broadening of the festival's identity over the preservation of dominant traditions that function to exclude them. In its present form, it risks preserving a traditional sense of community and locality at the expense of fostering more universal goals of equality, rights, and justice. (Jeong & Santos, 2004: 653–654)

Nor should tourism be treated as something outside (local) culture that brings in unfortunate and unnecessary changes. In fact, tourism is a part of culture. Bruner (1991) described effectively how tourism may become part of a culture, influencing and shaping it:

> A major reaction of the New Guineans to the encounter with tourists is amazement at the tourists' wealth and power. It is a momentous moment ... They realize that if they had more money, they too could go on the tourist ship, and they too could travel to distant lands to see other peoples. ... the touristic encounter not only raises questions about their current status, but it causes a revision of their conception of their own past, and it shapes their future ... They realize that the future will be in the direction of the Western culture, but they do not yet see clearly the path they will eventually follow. (Bruner, 1991: 245)

Regaining the *Genius Loci* of Places

As Saarinen *et al.* (2017) argued, tourism is a social and economic phenomenon which calls for proactive measures to help ensure positive development trajectories. The 'Genius Loci' project (2019), co-founded by a number of European Union countries, is intended to spotlight the heritage of small and medium-sized enterprises, traditional crafts and industries in order to revalue them and enhance their appreciation by tourists and the general public. The aim is for the tourist product to combine a number of attractions, such as museums, living industries and industrial heritage sites. The project aims to foster the tourist market in relation to three sectors in particular: fermented drinks (such as wine), the distilling process and their transformation into spirits; the heritage of the clay processing industries (pottery and brickwork); and traditional textile crafts (weaving) and the production of traditional textile fibres. Similarly, as Hunt (2018) reported, the city of Amsterdam has made significant marketing investments to shake off its reputation as a citywide red-light district and bolster its reputation as an arts and culture destination.

The following suggestions have been made to help destinations maintain or even regain their *genius loci*. (Note: The following general suggestions are taken from Christou *et al*., 2019c: 29. Even so, examples taken from other studies are also included in the points that follow – these are referenced.)

(a) Enforce and implement strict legislation to protect the destination's cultural and natural environment. For instance, in 2018 the Mayor of Santorini limited the number of cruise ship passengers disembarking to 8000 per day. The Greek island is only 76 km^2 in size, but it hosts millions of overnight stays (Smith, 2018).

We can't have small islands, with small communities, hosting 1 million tourists over a few months. There's a danger of the infrastructure not being prepared, of it all becoming a huge boomerang if we only focus on numbers and don't look at developing a more sustainable model of tourism. (Chrysogelos, an environmentalist, in Smith, 2018)

(b) Identify significant sites within the destination (that is, sites that reflect the character of the place, its culture and 'soul', such as spiritual sites/monasteries), and protect them. These sites may undergo excessive change as a result of tourism-related development and (unfortunately) of unruly visitor behaviour. In this regard, Petzet (2008) has questioned whether we are actually preserving the meanings of places and their 'spiritual message' which have been entrusted to us for a short time. In 2018, a remorseful tourist returned stones taken from the ancient Acropolis site many years before: 'I am sorry. I took these from a trail on the Acropolis in Greece many years ago. Please return them' (anonymous tourist, in Ekathimerini, 2018). In Australia, Uluru is a sacred site for the Aboriginal people, who repeatedly asked for it to be closed to the hundreds of thousands of tourists who climb the rock every year. They requested that visitors pay respect to the rock's significance (Ruck, 2012). Eventually, climbing on the popular 'Big Red Rock' was banned out of respect for its ties to Aboriginal culture (Hallinan, 2018a). More generally, authorities and site management bodies are urged to be extremely cautious if they see that a site is being rapidly or slowly transformed into a popular tourist attraction. Such sites may be fragile and risk being transformed from places of worship into tourist attractions, thereby having their identity altered and their *genius loci* obliterated. Sacred places such as chapels, abbeys, monasteries, temples and cathedrals have in some cases been converted into facilities for tourists, including hotels, restaurants and bars (see Newell-Hanson, 2018, for some examples). Even where these conversions are characterised by careful restoration of the building and sophisticated design, it is evident that the character and sense of the place has been altered: what was once a sacred site is now a hospitality venue.

(c) Protect and enhance the physical, natural and spiritual characteristics of these sites. For example, include plants that are linked to the destination's endemic flora. The role of tourists is also important for the protection and care of the destination. Walker and Moscardo (2016) suggested that it is important to use values as a basis for the design and delivery of experiences that encourage tourists to move beyond a sense of place to 'a sense of care of place'.

(d) Enrich the sensory environment of the sites and the social, psychological and spiritual experiences of visitors. Examples include the promotion of holistic experiences (giving visitors the opportunity to listen to hymns while at religious sites) and employing locals who want to share their knowledge with visitors at the site. Unwanted conflicts between locals and visitors as a result of 'mass tourism' and unruly tourist behaviour should be avoided.

(e) Extend the site's *genius loci* to the whole destination or region. According to Thompson (2003), the *genius loci* has practical value since it can help us to adopt more sensitive approaches to planning and construction. However, extending the *genius loci* of a particular site to surrounding areas can be a challenging task, especially in urban settings, given that 'modern' or 'foreign-imposed' architectural influences may already have altered the identity of the place. As the Norwegian architectural historian and critic Norberg-Schulz (1979) stated:

> In town, 'foreign' meanings meet the local genius, and create a more complex system of meanings. The urban genius is never merely local; although the examples of Prague, Khartoum and Rome have taught us that the local character plays a decisive role in giving the settlement its particular identity. (Norberg-Schulz, 1979: 32)

Nevertheless, destinations may protect the site's surrounding environment and, if possible, redesign it to reflect the site's character. In this way, the *genius loci* will be extended from the 'capsule' site into the adjacent environment. Micro-environments (sites) in a destination may not be affected by the overall changes that the place has experienced. Such sites may act as shielding 'capsules', since they manage to keep their identity, architecture and soul through the passage of time, possibly for hundreds of years. Other suggestions include: inviting local artists to include in their works elements derived from the particular site; creating local memorabilia that represent the destination's *genius loci*, thereby promoting local heritage and supporting positive (memory) linkage for visitors; and including 'references' to the site in the whole region, for instance, by including replicas of the site's objects in the surrounding areas.

A Diagram Illustrating 'Place' within the Context of Tourism

Along with the concepts of 'sense of place', *genius loci* and the transformation of places enable the construction of a diagram (see Figure 7.1) that captures the importance of 'place' within the context of tourism. The column on the left of the diagram represents the significance of places that are not just geographic locations but an amalgam of destination qualities (Smith, 2015). These are centres where biological needs are satisfied (Tuan, 1979); they evoke different emotions in people (Meyer, 2003) and may lead to strong attachment (Jepson & Sharpley, 2015) and even to topophilia (Tuan, 1990). The sense of place is created by the specific qualities possessed by a place that make it unique, with a strong 'identity' (Bloom, 1990), and by people's perceptions and interpretations of these qualities (Hausmann *et al.*, 2016).

At the heart of the concept of 'sense of place' lies the long-honoured notion of *genius loci* (Gayley, 1893; Lecompte *et al.*, 2017). Within a place, specific sites retain strong characteristics that are linked directly to the roots, heritage and above all the spirituality of the place and its people. These sites have managed to protect their soul, spirit and aura from time and from other impacts such as modernity and development.

It is unfortunate that places around the world may experience dramatic change due to tourism development. Overtourism, rapid and unplanned tourism-linked development and resort construction have caused serious degradation of landscapes and impacts on the environment, resulting in places being significantly changed (Diamond & Olito, 2019; Haines, 2016; Lange, 2015; Street, 2018; White, 2019). Places may even change beyond recognition, becoming 'symbols of ugliness' (Der Spiegel, 2018). They lose their character (Plog, 2001), the attractions within them lose their iconicity (Weaver & Lawton, 2007; Weidenfeld, 2010), and cheap souvenir stalls replace traditional markets (Diamond & Olito, 2019). Tourism may largely be blamed for this 'placelessness', which causes a place to lose its essence, identity and *genius loci*, giving way to standardised shops, resorts and landscapes (Relph, 1976; Shim & Santos, 2014; Terkenli, 2002) that are 'tourist-friendly' (Bates, 2012).

A number of academics and environmentalists have called for measures to protect destinations and their 'sense'. For instance, Petzet (2008)

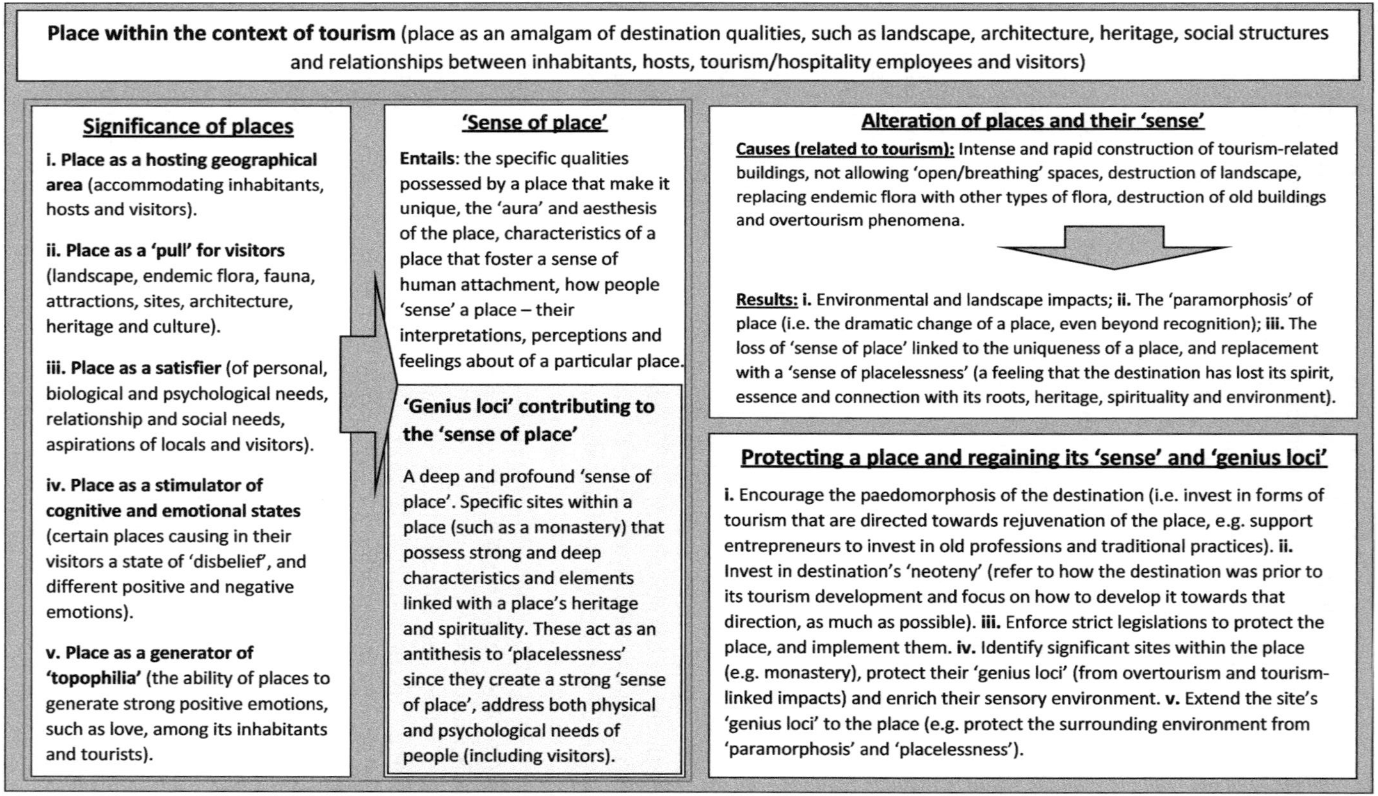

Place within the context of tourism (place as an amalgam of destination qualities, such as landscape, architecture, heritage, social structures and relationships between inhabitants, hosts, tourism/hospitality employees and visitors)

Significance of places

i. Place as a hosting geographical area (accommodating inhabitants, hosts and visitors).

ii. Place as a 'pull' for visitors (landscape, endemic flora, fauna, attractions, sites, architecture, heritage and culture).

iii. Place as a satisfier (of personal, biological and psychological needs, relationship and social needs, aspirations of locals and visitors).

iv. Place as a stimulator of cognitive and emotional states (certain places causing in their visitors a state of 'disbelief', and different positive and negative emotions).

v. Place as a generator of 'topophilia' (the ability of places to generate strong positive emotions, such as love, among its inhabitants and tourists).

'Sense of place'

Entails: the specific qualities possessed by a place that make it unique, the 'aura' and aesthesis of the place, characteristics of a place that foster a sense of human attachment, how people 'sense' a place – their interpretations, perceptions and feelings about of a particular place.

'Genius loci' contributing to the 'sense of place'

A deep and profound 'sense of place'. Specific sites within a place (such as a monastery) that possess strong and deep characteristics and elements linked with a place's heritage and spirituality. These act as an antithesis to 'placelessness' since they create a strong 'sense of place', address both physical and psychological needs of people (including visitors).

Alteration of places and their 'sense'

Causes (related to tourism): Intense and rapid construction of tourism-related buildings, not allowing 'open/breathing' spaces, destruction of landscape, replacing endemic flora with other types of flora, destruction of old buildings and overtourism phenomena.

Results: i. Environmental and landscape impacts; **ii.** The 'paramorphosis' of place (i.e. the dramatic change of a place, even beyond recognition); **iii.** The loss of 'sense of place' linked to the uniqueness of a place, and replacement with a 'sense of placelessness' (a feeling that the destination has lost its spirit, essence and connection with its roots, heritage, spirituality and environment).

Protecting a place and regaining its 'sense' and 'genius loci'

i. Encourage the paedomorphosis of the destination (i.e. invest in forms of tourism that are directed towards rejuvenation of the place, e.g. support entrepreneurs to invest in old professions and traditional practices). **ii.** Invest in destination's 'neoteny' (refer to how the destination was prior to its tourism development and focus on how to develop it towards that direction, as much as possible). **iii.** Enforce strict legislations to protect the place, and implement them. **iv.** Identify significant sites within the place (e.g. monastery), protect their 'genius loci' (from overtourism and tourism-linked impacts) and enrich their sensory environment. **v.** Extend the site's 'genius loci' to the place (e.g. protect the surrounding environment from 'paramorphosis' and 'placelessness').

Figure 7.1 Place in the context of tourism
Source: Author.

questioned whether we are truly preserving the meaning of places and their 'spiritual message'. Measures and practices identified include: controlling the number of tourists and their behaviour; curbing tourism-related (and often hitherto uncontrolled) development; providing financial support for heritage sites and research and education about site preservation and protection (Hall *et al.*, 2016); promoting a mentality of 'a sense of care of place' (Walker & Moscardo, 2016); and extending a site's *genius loci* to the surrounding area by protecting its heritage and architectural appearance.

Christou (2019b) recommended that destinations invest in their 'neoteny' and 'paedomorphosis'. The term neoteny was coined by Kollmann (1885) from the Greek words *neos* (young) and *teinein* (to extend). Within the tourism context, it refers to sustainable initiatives based on principles that reflect and extend the infant stages of tourism development. For instance, destinations are encouraged to introduce laws, regulations and strict policies that slow the rapid development of tourism. They should maintain open spaces, especially in areas of intense tourism, and protect the destination's natural landscape. They should also create obstacles to excessive change to a destination's culture in order to preserve the destination's intangible heritage and culture and to safeguard the 'spirit' of a place. This may be achieved by protecting sacred places and spiritual sites within the place from the impacts of tourism and tourist activity.

The other term, 'paedomorphosis', literally means to retain child-like characteristics when adult (some adult humans may display neotenic body features that other adult primates do not). Within a commercialised world, consumers develop affective attitudes towards child-like characteristics (Hellén & Sääksjärvi, 2013; Miesler *et al.*, 2011), and neoteny is manifested in juvenile products, such as cute items in souvenir shops and guest rooms. In the tourism context, however, paedomorphosis refers to maintaining or returning to earlier, more rejuvenating forms of tourism.

From this perspective, tourism stakeholders and locals should try in the first instance to retain the characteristics of their place, avoiding deformation and excessive change caused by tourism development. However, if the destination's landscape and environment do undergo change, then local stakeholders should try to regain the place's natural and cultural characteristics. For instance, governments should encourage and give financial support to entrepreneurs to invest in traditional professions and practices. Local authorities should promote activities that connect people (both locals and visitors) with the destination's heritage. The destination itself should promote and invest in forms of tourism that connect visitors with their inner child, such as nostalgia tourism and nostalgic references in accommodation and restaurant venues.

8 Giving and Receiving Places: Spiritual Tourism and Dark Tourism

Spiritual Sites and Spiritual Tourism

The difference between spiritual and religious tourism

'Both spirituality and religion are complex phenomena, multidimensional in nature, and any single definition is likely to reflect a limited perspective or interest' (Hill *et al.*, 2000: 52). Spirituality and religion may be intertwined, yet there are distinctions between the two. Although there is no scholarly consensus as to what precisely constitutes a religion, it may be regarded as a social-cultural system of ethics, morals, practices and worldviews that relates humanity to a particular supernatural and prodigious power.

Spirituality covers a broader spectrum of themes of which religion is one. The word derives from the Latin *spiritus* (soul) and is used to translate the Greek *pneuma* and Hebrew *ruach*; it was used within early Christianity to refer to the life of a person oriented towards the Holy Spirit. Spirituality may be explained as a process of personal transformation in order to realise one's search for meaning in life or one's fundamental principle. Recently, spirituality has acquired a much wider meaning, and it now includes wellness and self-improvement (Kato & Progano, 2017). Even so, it may still be regarded as a process of re-formation with the aim of recovering the original shape of man (that is, the image of God). This re-formation uses a mould that preserves the original shape: in Christianity there is Christ, in Buddhism there is Buddha and in Islam there is Muhammad.

Spiritual tourism explores the elements that contribute to the life of the body, mind and spirit, although there may or may not be a connection to religion. Cheer *et al.* (2017) noted both religious and secular motivations for spiritual tourism; the predominant secular motives include wellness, adventure and recreation. Prayag *et al.* (2016) found that tourists' encounters with *ayahuasca* (a hallucinogenic beverage traditionally consumed in Iquitos, Peru for spiritual and health purposes) were perceived as spiritual experiences that gave them a better understanding of themselves and others.

Religious tourism involves visiting a religious site or destination, with the primary focus being engaging with or intensifying a specific faith. The tourism and hospitality industry addresses the needs of particular tourists to abide by the rules of a specific religion. For instance, halal tourism caters for Muslims who abide by the rules of Islam, involving objects and actions that are permitted by Islamic teaching (Battour & Imail, 2016). In Malaysia, for example, hotels may choose not to serve alcohol and to have separate spa facilities for men and women. More generally, the provision of halal food may be considered as a competitive advantage for destinations that are positioned as Muslim friendly (Henderson, 2016).

People seeking spiritual experiences

Traumatic experiences can lead to a deepening of religion or spirituality (Shaw *et al.*, 2005) and the need for someone to heal their body or soul may underpin a travel quest to a spiritual centre, a sacred site or a spiritual person.

> Modern medicine can successfully treat many ailments that once took pilgrims to shrines, but the unresolvable human crises – hunger, poverty, loneliness, unemployment, and bereavement – continue to plague humanity. Belief remains the key element identifying the journey on the pilgrim-tourist path, and the motive can be sacred or secular. (Smith, 1992: 14)

Cheer *et al.* (2017) suggested that if spirituality is the goal, travelling seems like an ideal setting within which to seek and find it. According to Willson (2011), travel to spiritual centres can assist in a quest for inner peace, since it acts as an instrument for healing damage caused by secular and materialistic environments. Individuals may travel to particular places in order to make contact and have deep and meaningful conversations with spiritual people, as in the following example:

> Descending the once narrow path – which had become wider – visitors came across wooden arrows indicating: 'To Father Paisios.' [*At Mount Athos*] ... a familiar sign on the fence ... read: 'Write what you want and put it in the box, and I will help you more with prayer. I will therefore have more time to be able to help even more people who are suffering.' Next to the sign was a glass jar with pencils and paper, as well as a mailbox in which to place their notes. Yet, it was not ever sufficient for anyone to simply leave a note; everyone wanted to meet the man of God ... Everyone waited patiently – sometimes for hours ... There were also many who were impatient as to when the Elder would finally appear. If he delayed for a long time, they started to shout, 'Elder, please come! We are in need of you. We are in need, Elder!' (Holy Hesychasterion, 2018: 395–396)

A number of studies have examined what motivates people to engage in a spiritual or religious trip. Specific types of tourism have been shaped by the need to engage in particular forms of spiritual travel activity, such as 'Ashram tourism' (Sharpley & Sundaram, 2005) and 'halal tourism' (Battour & Ismail, 2016). Andriotis (2009: 72–79) found that visitors' motivations for visiting Mount Athos in Greece vary, but that the different activities and behaviours are characterised by certain core elements:

(a) *Spiritual element.* A large number of visitors to Mount Athos are pilgrims (referred to in the study as 'proskinites'), motivated by authentic Orthodox Christian tradition and faith. While in the church, pilgrims express their devotion to God by praying, making the sign of the cross and lighting candles. Mount Athos allows pilgrims to meditate away from the cares and distractions of their everyday lives.

(b) *Cultural element.* Mount Athos is a living museum of history and art, and experience of culture is a major element for most visitors interested in Byzantine architecture or the monastic life. Seeking authentic experiences is also a driving motivator for visitors who refer to Mount Athos as 'real', 'genuine' and 'untouched'.

(c) *Secular element.* The communal nature of accommodation and life at Mount Athos encourages social interaction between visitors, who adopt the communal lifestyle by being more open/talkative to strangers than they would be in their home settings. The act of taking photographs is an important social activity for the travellers which also strengthens bonds between them.

(d) *Environmental element.* The place itself combines spiritual quest with physical journey. Visitors kitted out for trekking the Himalayas have also been found trekking on Mount Athos.

(e) *Educational element.* Students were found to be motivated to visit Mount Athos after attending lectures on Byzantine art. Visits to Mount Athos for cognitive purposes were also evident, with visits offering the potential for cathartic experiences related to learning. Elders may establish interpersonal relationships with visitors by disseminating (spiritual) knowledge to them, and visitors may educate themselves through conversations with highly educated monks.

Hyde and Harman (2011) examined the motives of Australians and New Zealanders for a secular pilgrimage to the Gallipoli battlefields in Turkey. Five distinct motives were identified: spiritual, nationalistic, family, friendship and travel. Spiritual tourism can of course be combined with other activities, such as walking in the case of the Nakahechi pilgrimage trail to Kimano in Wakayama, Japan (Kato & Progano, 2017). In fact, Lois-González and Santos (2015) suggested that tourists often combine pilgrimage motivations such as displacement for religious purposes, with tourist motivations such as the need to escape from the pressures of daily life, the search for various landscapes and the need to mentally unwind. 'Retreat tourism', a specialty subsector of wellness tourism, may provide opportunities for people to engage in a combination of activities, practices and treatments that aim to balance the body, mind and spirit (Kelly, 2012).

Places as spiritual centres for visitors

Religious tourism is one of the oldest forms of tourism activity (Matheson *et al.*, 2014). From ancient times, particular places of religious importance (such as the temples of Karnak in Egypt, Delphi in Greece and the Pantheon in Rome) have attracted visitors. Jerusalem is considered holy to the three Abrahamic religions of Judaism, Christianity and Islam. Christianity reveres Jerusalem both for its Old Testament history (that is, as the place of prophets such as David, Ezekiel and Jeremiah) and for its New Testament history (that is, as the place where Jesus Christ was crucified and resurrected). The church of the Holy Sepulchre, also referred to as the Church of Resurrection, is of particular importance for faithful Christian visitors. The Dome of the Rock shrine is a significant Islamic site and one of the most recognisable landmarks of the city, and the Western Wall attracts millions of visitors:

> The Rabbi of the Western Wall ... noted that the masses who visited the Western Wall [*approximately 2.5 million people during the Hebrew month of Tishrei*] served as impressive and inspiring evidence of the unifying connection we all have with the Western Wall. (Western Wall Heritage Foundation, 2019)

Shackley (2001: 2) classified sacred sites into 11 categories:

(a) single nodal feature (such as Canterbury Cathedral or Hagia Sophia, Istanbul);
(b) archaeological site (such as Machu Picchu, Peru);
(c) burial site (such as the Pyramids at Giza);
(d) detached temple (such as Angkor Wat);
(e) whole town (such as Varanasi, India and Jerusalem);
(f) shrine or temple complex (such as Catherine's Monastery, Egypt);
(g) 'Earth energy' site (such as the Nazca Lines, Peru);
(h) sacred mountain (such as Uluru in Australia, Mount Athos in Greece and Mount Fuji in Japan);
(i) sacred islands (such as Mont Saint-Michel, France);
(j) pilgrimage focus (such as Mecca, Saudi Arabia); and
(k) secular pilgrimage (such as Holocaust sites).

Of particular interest are the various monastic communities that attract religious (or non-religious) visitors. Mount Athos is one of the most important centres of Christian Orthodox monasticism (see Photo 8.1). It is home to 20 monasteries that follow the coenobitic system. There are also 'cells', usually houses with a small church where a small number (usually between one and three) of monks live, under the spiritual supervision of a monastery. In a similar case, Meteora is a rock formation in central Greece which hosts a number of monasteries that are included on the UNESCO list of World Heritage Sites. Along with their spiritual heritage, the landscape and stunning sceneries of both these sites play a significant role in the pilgrims' experience (Della Dora, 2012).

> I consider myself fortunate to have met in the course of my life remarkable people such as Elder Paisios. They have been instrumental in helping me find my way. Through the wondrous experiences they graciously offered me, they provided answers to disquieting questions I had been grappling with for many years: Is there any meaning to the universe? Does God exist? How should I live? Who am I, and what is my inner nature? Will I cease to exist after death? (Farasiotis, 2005: Preface)

Photo 8.1 Mount Athos, Greece. The quest for what is spiritual, religious and meaningful in life has always triggered people's curiosity. Mount Athos, or Holy Mountain (*Agion Oros*), a mountainous peninsula in north Greece, is home to 20 monasteries. Prayers take place at night, with services starting at 2am and finishing at 6am. Monks spend much of the day working, cleaning the guest houses for visitors and preparing dinners. Visitor numbers are controlled so that tranquillity is maintained. Visitors may stay in the monasteries provided that they apply in advance for the relevant permit, a *diamonitirion*. Hospitality is provided free of charge, with bed and board offered to permit holders. The photo shows Saint Panteleimon Monastery, which has historical and liturgical ties to the Russian Orthodox Church
Source: Author.

Nevertheless, it should be emphasised that spiritual places and sites may not necessarily act as a pull of attraction only for visitors who seek spiritual awakening and enlightenment. Wong *et al.* (2013), who targeted respondents visiting a Buddhist site, identified two types of visitors: 'sightseers' and 'cultural/heritage' visitors. Similarly, Bond *et al.* (2015) found that visitors to important religious sites (including Canterbury Cathedral and the Shrine of Our Lady of Walsingham) reported a high level of interest in history and culture, with cathedral visitors showing little interest in personal spiritual benefits.

Spiritual, religious and pilgrimage sites continue to provide fertile ground for in-depth explorations and investigations of tourism (see Belhassen & Ebel, 2009; Belhassen *et al.*, 2018; Bond *et al.*, 2015; Gill *et al.*, 2019; Shuo *et al.*, 2009), and in all likelihood will continue to do so in the future.

Thanatourism and Places of Atrocity

Although genocide camps, cemeteries, battlefields and assassination locations are primarily associated with death and grief, they are also among the world's most visited tourist locations (Johnston, 2015) (see Photo 8.2). Examples include the memorial and museum at Auschwitz-Birkenau (Poland), the National September 11 Memorial and Museum (New York), the Hiroshima Peace Memorial Museum (Japan), Chernobyl (Ukraine) and the Murambi Genocide Memorial Centre (Rwanda).

Freud (1984) posited that the death instinct of 'Thanatos' (death) operates from the beginning of life in opposition to the life instinct of 'Eros' (sex/life). In Greek mythology, Thanatos was the personification of death: when humans died, he carried them to the underworld. Hesiod (a Greek poet active around 750 BC) considered Thanatos to be the son of Nyx (Night) and Erebos (Darkness) and the twin of Hypnos (Sleep). His association with 'dark' notions can be explained in terms of the ancient Greek belief in Hades, the god of the dead and the king of the underworld. The kingdom of Hades was depicted as a misty and gloomy abode where all mortals went when they died.

Certain religions (such as the Abrahamic faiths) believe in an afterlife, that is, in the soul's existence after a person's physical death. From their perspective, the death of a body may not necessarily lead a soul to a 'dark' or punishment place such as the Jahannam of Islam. Instead, a soul may be placed somewhere 'brighter' and better than this world, reaching a state in which peace, justice and endless joy prevail. Ancient Christian teachings dating from the 1st century CE consider in great detail the spiritual world beyond death (see Rose, 2009).

The term 'thanatourism' is closely associated and often used interchangeably with terms such as 'dark tourism' and 'grief tourism', although there remains some debate about what each term entails (Friedrich &

Photo 8.2 Old Jewish Cemetery, Prague, Czech Republic. Around 12,000 tombstones are crammed into the Old Jewish Cemetery, which is an important historical monument and popular site in Prague. Space was scarce, hence bodies were buried on top of each other, with graves up to 10 bodies deep
Source: Author.

Johnston, 2013). Light (2017) noted that two decades of research have not convincingly demonstrated that 'thanatourism' and 'dark tourism' are distinct forms of tourism, and both appear in many ways to be little different from 'heritage tourism'. Nonetheless, Seaton (1996) claimed that thanatourism is distinctive:

> Thanatourism is travel to a location wholly, or partially, motivated by the desire for actual or symbolic encounters with death, particularly, but not exclusively, violent death, which may, to a varying degree be activated by the person-specific features of those whose deaths are its focal objects. (Seaton, 1996: 240)

According to Seaton (1996), thanatourism includes five distinct travel activities:

(a) Travel to witness public enactments of death. In the past, this was in the form of gladiatorial combats to the death, and Christian martyrdoms in Rome. Now it may take the form of people slowing down their cars to gaze at motorway pile-ups or rushing to disaster scenes.
(b) Travel to see the sites of mass or individual deaths. Mass death sites include Auschwitz or the Colosseum in Rome, and individual death

sites include the room where the Princes in the Tower of London were murdered.
(c) Travel to internment sites of, and memorials to, the dead. This kind of thanatourism may include visits to crypts, war memorials, catacombs and graveyards, such as the Jewish Cemetery in Prague. There are also hundreds of battlefield sites in England and Scotland, although very few of these have adequate infrastructure or any form of interpretation to attract visitors (Miles, 2014).
(d) Travel to view material evidence, or symbolic representations, of death in locations unconnected with their occurrence. This includes museums that display murder weapons or the clothing of murder victims.
(e) Travel for re-enactments or simulation of death. This includes religious presentations that restage the death of Christ such as the Passion Play at Oberammergau in the German Alps, and battle re-enactments.

With specific reference to contemporary Asia, Cohen (2018) distinguished two main varieties of thanatourist phenomena that draw ancestor worship in particular. First, customs of ancestor worship at sites of deceased non-kin persons (especially of celebrities) mainly attract domestic visitors. Secondly, traditions more loosely and selectively associated with commemoration and worship at recent historical sites of massive death (for instance, battlefields, and places where atrocities were committed or disasters occurred) attract both domestic and international visitors.

Elsewhere in the literature, there are references to 'genocide tourism', as in the case of the mass slaughter in Rwanda in 1994 (Friedrich & Johnston, 2013), 'disaster tourism' (Robbie, 2008), 'grief tourism' (Dunkley *et al.*, 2007), 'suicide tourism' (Miller & Gonzalez, 2013) and 'Holocaust tourism', which emphasises the meaning, value and extent of visits to specific places that honour the victims of Nazism (Hartmann, 2014).

Stone (2006) clarified certain aspects of the supply side of dark tourism with a spectrum ranging from 'darkest' to 'lightest'. Within this framework, he outlined seven types of dark suppliers: dark camps of genocide, dark conflict sites, dark shrines, dark resting places, dark dungeons, dark exhibitions and dark fun factories. Sites merely 'associated' with death and suffering occupy the 'lighter' end of the dark tourism spectrum, while the 'darker' end of the spectrum includes sites of actual death and suffering.

Thanatourism, travel motivation and the tourist experience

Although human fascination with death is a constant, thanatourism has expanded significantly, partly owing to the influence of the media

(Knudsen, 2011), becoming significantly more common within the last few decades. Tourist fascination with sites associated with death and tragedy has received considerable academic attention (see Dunkley *et al.*, 2007; Friedrich & Johnston, 2013; Light, 2017; Sharpley, 2005; Tanaś, 2014). People may be driven to particular macabre exhibitions and museums out of curiosity and interest in death (Johnston, 2015), as with entertainment-based museums of torture. Best (2007: 38) characterised dark tourists as 'individuals who are motivated primarily to experience the death and suffering of others for the purpose of enjoyment, pleasure and satisfaction'.

Other reasons have been proposed: a fascination with evil (Lennon, 2010), nostalgia (Tarlow, 2005), pedagogical and educational purposes (Cohen, 2011), interest in genealogy and family history (Boyles, 2005; Buntman, 2008) and cultural interests (Farmaki & Antoniou, 2017). Thanatourism has also been viewed as a means of producing 'authentic feelings' (Knudsen & Waade, 2010); hence, going to thanatouristic sites is a way for people to trigger certain emotions, like watching a melodrama in order to cry (Knudsen, 2011). 'Dark' emotions, such as pain, horror and sadness, may be part of a tourist's experience at dark sites (Ashworth, 2008). At Culloden battlefield in the UK, a 360-degree visually engaging film of the battle provides insight into the emotional reality of the conflict:

> You were able to imagine the noise, sights and smells of the battle. The guns, people screaming, [the] horses, [the] smell of blood [and] gunsmoke. (Female visitor aged 50–65, in Miles, 2014)

In other cases, a visit to a Holocaust site such as Auschwitz may represent a form of 'pilgrimage' tourism (Ioannides & Ioannides, 2006). Slade (2003) found that Australians and New Zealanders were motivated by 'nationhood' to visit the battlefield of Gallipoli in Turkey; that is, they were visiting the place where one of their great nation-building stories is set.

In his typology of dark tourism/tourists, Sharpley (2005) integrated both supply and demand. His model recognises the heterogeneity of dark tourism demand and supply: not all so-called 'dark tourism attractions' are intended to be 'attractions', and not all tourists who visit these attractions are strongly interested in death. Similarly, Raine (2013) identified different categories of visitors at three burial grounds, ranging from 'darkest' to 'lightest' tourists. Stone and Sharpley even argued that dark tourism may have more to do with life than with the dead and dying:

> [C]onsuming dark tourism can help individuals, within a social structure, to address issues of personal meaningfulness – a key to reality, thus to life and sustaining social order, and ultimately to the maintenance and continuity of ontological security and overall well-being. (Stone & Sharpley, 2008: 590)

Current Ethical Issues and Concerns Related to Spiritual and Atrocity Places

Authenticity and the recreation of sacred/atrocity experiences for tourist consumption

An Indonesian museum created and displayed a wax statue of Adolf Hitler in a dominating pose against a backdrop of the Auschwitz concentration camp. The statue was a popular attraction for visitors, who

Photo 8.3 'Nea Moni', UNESCO World Heritage Site, Island of Chios, Greece. The display of skulls at Nea Moni heritage and religious site on the island of Chios (in the Aegean Sea) may trigger emotional reactions in visitors, such as sadness. It also gives rise to a number of ethical issues. Such quandaries include: (a) whether this display pays tribute and honour to the deceased who have died in a battle; (b) whether people's bones and skulls should be displayed for the gaze of others; (c) whether visitors should be exposed in this way to the dramatic slaughters that occurred on the island; and (d) whether these and similar displays and narratives trigger aggression and hatred
Source: Author.

took selfies in front of the exhibit; however, it sparked global outrage and drew condemnation from campaign groups, foreign media and human rights organisations. After the exhibit was removed, the museum manager said:

> We don't want to attract outrage ... Our purpose to display the Hitler figure in the museum is to educate [people]. (Deutsche Welle, 2017a)

A number of studies have examined the linkage of 'authenticity' and tourism/tourist experiences (see, for example, Belhassen *et al.*, 2008; Cohen, 1979, 1988; Moscardo & Pearce, 1986). The recreation of sacred sites and the staging of divine experiences raises particular issues related to the authenticity and genuineness of spiritual experiences. According to Blackwell (2007), there is a growing interest in developing purpose-built religious tourism attractions on non-sacred sites, such as the Holy Land Experience in Orlando, Florida. In effect, this religious theme park is highly desirable for the home population, since it can easily be visited without the inconvenience of a long journey or exposure to environmental or political danger. In this connection, Collins-Kreiner *et al.* (2017) provided an extensive account of pilgrimage, and more specifically Christian tourism, to the Holy Land during a security crisis.

There is ongoing debate within the academic community regarding what sites, locations or experiences may be labelled as 'authentic' (see Cohen, 2011), and to what extent 'authenticity' should be presented at thanato-places or other dark sites. For instance, one of the most contentious issues in constructing the US Holocaust Museum was the debate over whether to display the actual hair of the victims (Linenthal, 2001).

Photo 8.3 provides an example of a heritage site that may give rise to a number of ethical dilemmas.

Overtourism combined with unethical and vulgar tourist behaviour at specific places

Specific sacred places around the world are often called on to address the challenge of overtourism (see Photo 8.4) although, rather ironically, the COVID-19 coronavirus outbreak has contributed to its elimination within a very short period of time (BBC, 2020; Legorano, 2020). Despite this, overtourism is frequently (yet not necessarily) combined with inappropriate and unethical behaviour on the part of visitors. Angkor Wat in Cambodia has ties to both Hinduism and Buddhism. The temple complex is a vastly popular tourist attraction, with tourists lining up for tickets as early as 4.30am (Hallinan, 2018a).

In Florence in 2017, the mayor said that he valued tourism but also wanted the city to be respected; therefore, he introduced the regular hosing down of the steps of the Santa Croce Basilica:

Photo 8.4 Piazza di San Pietro, Vatican City. Thousands of visitors come to Piazza di San Pietro in Vatican City, with large queues of visitors waiting to enter St Peter's Basilica. The prevention of overcrowding and mass tourist phenomena at popular sites remains a challenge for many destinations, although rather ironically and sadly the Coronavirus did stop inflows of tourists in many countries
Source: Author.

People don't want to see empty bottles and greasy church steps. If we get a reputation for being rubbish-strewn, we lose the quality tourists. We want to put people off from camping out. If they sit down, they'll get wet. Instead of imposing fines, we thought this measure was more elegant. (Ahluwalia, 2017)

Van Tilburg, the founder of the Easter Island Statue Project, reported the following about the sacred stone anthropomorphic monoliths ('moai') there:

While the increased foot traffic alone has hindered conservation efforts, these days visitors to the sacred sites can behave in a vulgar, disrespectful fashion by trampling prohibited spaces, sitting on graves and climbing on the moai themselves in the service of getting pictures picking their noses. (Salisbury, 2019)

Taking selfies at sacred and atrocity places and sharing them on social platforms

Selfie-taking has been characterised as a new form of touristic looking, in which tourists themselves are the objects of the self-directed tourist gaze (Dinhopl & Gretzel, 2016). Taking selfies at horror sites is a growing trend on social networking sites (Hodalska, 2017), and taking and sharing selfies at sacred places and atrocity sites raises certain ethical issues. For instance, Sedlec Ossuary, the 'Church of Bones' chapel in the Czech Republic, draws hundreds of thousands of tourists annually. The holy site enforced stricter rules to curb a rise in inappropriate and derogatory photographs. Tourists were found taking insensitive selfies and manipulating bones to stage more interesting photos, thereby desecrating the holy site (Compton, 2019). In 2015, visitors to sacred sites in Cambodia were arrested and deported for taking nude photos there (National Post, 2015). Similarly, two US tourists were detained in Thailand after posing for a nude photo in front of a temple (Molloy, 2017). After a Czech man lifted up his girlfriend's skirt and splashed holy water on her bottom at a Hindu temple in Bali, outraged officials warned that tourists who disrespect sacred sites will be sent home or face 'purification rituals' (Webber, 2019).

According to Light (2017), actions like these are carried out by a small minority of tourists, and most visitors are deeply engaged with the places of death and suffering that they visit. Nevertheless, Sergei Loznitsa's documentary *Austerlitz* showed how casually some tourists treat Holocaust memorials:

> Nearly all of them are holding up their phones to take pictures ... Selfie sticks are adjusted at the entrance to the memorial, where the iron gate reads, 'Arbeit macht frei' (Work makes you free). Even here, visitors check in on Facebook and post their selfies to Instagram. (Deutsche Welle, 2017b)

In 2017 an Israeli-German writer copied from social media 12 selfies taken at the Berlin Holocaust memorial. In the background of the photos, he replaced the memorial with macabre scenes from concentration camps; young selfie-takers now appeared to be surrounded by emaciated bodies and corpses. The modified selfies were then presented on a website called 'Yolocaust', a combination of the words Holocaust and the popular social media hashtag Yolo ('you only live once'). People in the images were criticised for being 'disrespectful' and 'foolish' (Gunter, 2017).

The commodification of atrocity and sacred places

> [T]ouring places where memories of torture, dehumanization, and death are invited has an ethical obligation to interweave what may appear to some as banal signs of commodification with what otherwise might feel spectacularly tragic. (Bowman & Pezzullo, 2010: 198)

Tourism at places of death and suffering raises the ethical issue of whether it is acceptable to profit from macabre events and from people's deaths (Garcia, 2012; Light, 2017; Seaton, 2009). Commodification denotes the action of treating something as a mere commodity; in the context of tourism, commodification refers to using a place's culture and heritage in order to make profit. According to Olsen (2003), the commodification of religion and religious built heritage takes two main forms:

(a) Tourism puts pressure on sacred sites and customs. Although it has been argued that tourism provides opportunities to preserve religious sites and rituals, in many cases more tourists than pilgrims visit religious sites. This raises concerns over the conservation of these sites and their 'sense of place', and the risk of disturbing those who have come to the sacred place to worship. In many instances, religious sites and rites have been shaped to meet tourist expectations, with entrepreneurs taking items of religious significance and transforming them into religious souvenirs, thereby changing their original meaning or inventing new meanings for them. Christou et al. (2019c) revealed that the spiritual 'sense' of a particular site – in their case, a monastery – was kept almost unaltered because it managed to preserve its architecture, sensory environment with candles and smell coming from the 'thymiatos' (censer), opportunities for visitors to engage in conversations with spiritual people, and the aura of the place (with masses and religious ceremonies that visitors can attend). In the same study, the authors provided suggestions as to how a place can keep its 'sense' despite tourism pressures.

(b) Religious groups commodify their beliefs and doctrines for economic gain. It is common to find religious sites that have souvenir shops, and entry fees are often charged for sacred places and sites of worship such as churches. At St Nicholas Church, the famous Baroque church in Prague, a 'sightseeing' entrance fee is charged; according to the official site, payments can be made in cash or by credit card at the cash office of the church (Kostel Sv. Mikuláše, 2019). Rudgard (2017) reported that British cathedrals and churches have seen a drop in visitors, with entrance fees discouraging tourists. People may argue that places of worship should not charge an entrance fee, and such fees might be considered as unethical on the grounds that they force visitors to 'pay to pray'. A spokeswoman for St Paul's Cathedral in London (in Rudgard, 2017) expressed the following view:

As St Paul's receives little external funding, visitors who pay to enter for sightseeing remain absolutely crucial for us to be able to maintain both the fabric and rich, faith-led working life of the Cathedral.

Sacred places and site entrance constraints

It can be argued that placing constraints on entry to sacred sites may deprive visitors of the opportunity to have certain experiences. Even so, destinations set limits in order to protect the sacredness of their sites. In 2018, the Bali administration prepared a new regulation banning tourists from entering the most sacred parts of temples after a series of incidents in which badly behaved tourists desecrated holy sites on the island. In one such incident, a photo of a tourist sitting on top of a shrine went viral on different media platforms (Straits Times, 2018). The giant monolith of Uluru (otherwise known as Ayers Rock) is sacred to its indigenous custodians, the Anangu people, who have long implored tourists not to climb the sacred rock. Despite this, thousands of tourists rushed to climb Uluru before the activity was banned by the authorities in 2019:

> People right around the world … they just come and climb it. They've got no respect. (Rameth Thomas, in BBC, 2019)

> It's difficult see what that significance is … It's a rock. It's supposed to be climbed. (Visitor who climbed the rock, BBC, 2019)

> There were jeers from a small group of indigenous women. 'Get off the rock,' they shouted as two men from Germany – a father and son – made their way down. (BBC, 2019)

Elsewhere, Saudi Arabia decided in 2019 to open its doors for the first time to travellers who want to visit for non-religious reasons; however, the restriction that only Muslims can visit the holy cities of Mecca and Medina has been maintained (Mzezewa, 2019).

Dress codes in sacred places

Modesty in clothing is something that has long been urged in sacred places (such as monasteries and churches) and in specific states/countries. Qatar's Islamic Culture Centre launched initiatives to educate international visitors on dress codes. In 2014 the country launched a social media campaign urging tourists to dress 'modestly', asking people to cover their shoulders and knees in public and to respect the country's Islamic values (Dearden, 2014). In Vatican City, tourists entering St Peter's Basilica are required to dress modestly. In 2010 the Swiss Guards (the Pope's private army) appeared to apply these rules to the entire Vatican City State at a time when the Catholic Church was battling scandals over paedophilia, leading to visitors accusing the Vatican of 'hypocrisy' (Squires, 2010):

> Given all the scandals the Church has been involved in, what possible right can it have to be preaching about the morality of sleeveless dresses? (Maria, in Squires, 2010)

Nonetheless, throughout the millennia, spiritual leaders have emphasised a shift from over-attention on how to dress the body towards a focus on how to 'dress' the soul with virtues such as of kindness and benevolence:

> The soul of a true human being is beautiful, full of inner purity, and this beauty is not only present within the soul but is carried over into one's

appearance ... Father Tychon used to sew his own monastic caps using the cloth from worn out cassocks. The caps he made looked like bags, but when he wore them, they were full of grace. No matter how old or unkempt his clothes were, they looked beautiful because his soul had so much beauty ... Far greater is the worth of a single blessed man like Father Tychon, who changed his inner man and thus became holy in appearance, than the worth of so many others who care constantly putting on new clothes and yet keep wearing the same man inside, buried under layers of sin. (Elder Paisios, in Flesoras, 2011)

'Entertainment' and 'anaesthetisation' of visitors at sacred or atrocity places

According to Garcia (2012), the main issue with what are referred to as 'ghost tours' is the tension between educating visitors about true stories of people who lived and suffered in a particular place, on the one hand, and entertainment (as a commercial activity), on the other. However, this raises the ethical question of whether human suffering is to be offered to others for entertainment purposes. The challenges of managing tours of this type derive mainly from ethical and interpretative considerations.

Nonetheless, a number of destinations and dark sites offer tourists the opportunity to participate in a ghost tour. Port Arthur in Tasmania is a UNESCO World Heritage Site and one of Australia's favourite tourist destinations. More than 1000 people died at Port Arthur. The historic site offers a 90-minute lantern-lit ghost tour in which visitors walk through the buildings and ruins of the site and hear vivid stories about convicts and the site's past as a penal settlement (Port Arthur Historic Site, 2019).

Problematically, visitors may view certain sacred or atrocity sites as 'entertainment' places, and there is ethical debate as to how places associated with suffering or death should be presented to visitors. Light (2017) argued that the educational role of such places is compromised by an emphasis on entertainment and spectacle. Sharpley and Stone (2009: 111) called this 'dark edutainment', while Dale and Robinson (2011: 213) called it 'dartainment'.

It has also been argued that tourism at sites of atrocity may anesthetise rather than sensitise visitors, as increased contact with suffering, the macabre and 'horror' may make these things seem more normal and acceptable (Ashworth, 2008; Ashworth & Hartmann, 2005; Robb, 2009). Wright (2016) has even presented a futuristic scenario in which a wealthy elite would hunt and kill other humans as part of the tourism entertainment industry.

Exposure to danger in the quest of 'dark experiences'

Participation in any type of tourism activity comes with risks, and some tourists actively seek adventure, 'thrills' and the emotion of 'fear'

(Cater, 2006). Lepp and Gibson (2008) found that novelty-seeking tourists were less likely than organised mass tourists to perceive health issues and war as posing a risk to them as tourists. One English man took vacations in the war zones of Afghanistan, Iraq and Somalia:

> It's a risk and reward ... The reward is that you see more, and you learn more. We're not born to do nine to five jobs. You have to explore a little bit and try to understand the world you're living in ... They [*Somalians*] are loving people who do their best and put their life on the line for you. (Drury, in Monks, 2016)

Buda (2015) investigated how, in places where conflict is ongoing, certain tourists access the death drive in ways that confuse the binary oppositions of life/death and fun/fear:

> For Amru, travelling to Iraq also gave him a different understanding of life and death through his feelings of fear ... friends and people around me put[ting] me down sometimes. I mean, they would say 'Are you crazy?' and stuff. But I had this feeling that pushes me and said I wanted to go. (Quote derived from Buda, 2015: 46)

Current issues in travel euthanasia, assisted suicide and suicide places

Certain people may choose to commit suicide outside their proximal environment, often at a well-known location (Gross *et al.*, 2007) that carries the stigma of being a 'suicide spot/place'. Zhi *et al.* (2019) found that individuals who travel with the intention of committing suicide may choose locations with personal or cultural symbolism. Hence, these places are associated with high rates of suicide and may become popular attractions among tourists who are interested in taking their own lives there.

A forest in Japan (Aokigahara) has become infamous as a suicide spot (Salisbury, 2018). Signs on the trail paths promote a suicide hotline with the words, 'Life is a precious thing that your parents gave to you', and locals patrol the forest, talking to people who are alone or showing signs of depression and suicidal intent (Rich, 2018). In 2012, Australians paid tribute to Don Ritchie, who lived in an area of Sydney called 'the Gap', a beautiful yet notorious suicide spot, and who over 50 years saved 160 people (perhaps many more) from suicide, offering people conversation and a cup of tea, sometimes even physically restraining them from jumping into the sea (BBC, 2012).

Certain nations maintain a firm position against assisted suicide, but there are still documented cases of people travelling to specific locations to end their lives (Shondell Miller & Gonzalez, 2013). There is a difference between euthanasia and assisted suicide: euthanasia involves specific steps, and the 'final deed' is undertaken by someone other than the individual (such as a doctor); assisted suicide involves helping someone to take their own life themselves. In the Netherlands, both euthanasia and assisted suicide are legal, provided the patient is enduring unbearable suffering and there is no prospect of improvement (Davis, 2019). Although assisted suicide is strictly restricted in many countries, people may visit Switzerland for the sole purpose of committing suicide (Gauthier *et al.*, 2015), with right-to-die organisations in the country assisting their members to commit suicide (Ogden *et al.*, 2010). Despite this, Safyan (2011: 288) argued that 'death tourism as a new "industry" has revived international debate regarding the legislation of physician-assisted suicide and euthanasia'.

Closing Remarks

Travel for leisure, trade, education, or any other reason has accompanied humanity for millennia, but it has been – and in all likelihood will continue to be – severely affected by wars and crises. A global economic recession or a pandemic have proven to bring even the vast tourism industry to a complete standstill, but only for a certain period. Like a phoenix, though, tourism rises from the ashes, because it is fuelled by the seemingly endless need of humanity to explore, escape, unwind, and acquire additional knowledge, and the willingness of people to provide for those in need with a ride, a bed, or a meal.

In an era of numerous challenges for the tourism industry – particularly the COVID-19 pandemic, but also unethical behaviours and egocentric attitudes – this book aims to remind us of some of the core rudiments of the travel, tourism and hospitality domain. A more philosophical approach to tourism can help all of us understand that there is more to life than taking advantage of the natural environment, customers, employees, or hosts, for the sake of profiteering or self-indulgence. Therefore, the acquisition and channelling of certain rudiments and notions such as philanthropy are deemed crucial at a personal, organisational and societal level. The cultivation and circulation of virtues such as love, kindness, patience and charity in tourism are of the utmost importance if organisations are to be associated by their guests with terms such as 'anthropocentric', 'extraordinary', 'unexpected', 'quality' and 'satisfaction'. Professionalism and quality-driven service provision are highlighted in this book. However, such values should be enhanced by genuine care towards our guests, hence reflecting our ancestors' tactics of offering real care to those that are away from home or away from their safety and/or comfort zone. Our guests trust that we will convey them safely to their loved ones, or accommodate, feed and guide them, while taking care of their health and well-being. For them, a simple ride, a clean and comfortable room, or a nice dish is what they expect from us. What is perhaps less expected from us is to offer something free of charge, show empathy for something that they are worried about, or get out of our comfort zone to deal with a complaint that they have expressed. On the other side of the spectrum, tourism employees have encountered, and continue to face, enormous physical and psychological pressure as a result of unemployment, non-empathetic managers, rude or selfish customers, and harsh working conditions. They deserve to be treated with care and kindness by both their employers and their guests.

As a concluding note, severe crises like the COVID-19 pandemic allow us to appreciate what we take for granted, such as an international flight or even a meal at a local restaurant. Hopefully, such crises also lead us to reflect on our actions, such as our behaviour towards our employees, guests, or service providers. Sometimes, it seems as though there is a need for a crisis, or even a book such as this one, to remind us of the importance of acting in a gentle and kind manner.

References

Agapito, D., Valle, P. and Mendes, J. (2014) The sensory dimension of tourist experiences: Capturing meaningful sensory-informed themes in Southwest Portugal. *Tourism Management* 42, 224–237.

Ahluwalia, R. (2017) Florence church steps hosed down to deter tourists. *Independent*, 1 June. See https://www.independent.co.uk/travel/news-and-advice/florence-church-tourists-water-wet-hose-santa-croce-basilica-picnics-eating-food-italy-a7767111.html (accessed November 2019).

Airbnb (2019) *Hurricane Dorian – Southeast United States*. See https://www.airbnb.com/welcome/evacuees/dorianmainland (accessed September 2019).

Alexander, M., Chen, C., MacLaren, A. and O' Gorman, K. (2010) Love motels: Oriental phenomenon or emergent sector? *International Journal of Contemporary Hospitality Management* 22 (2), 194–208.

Alexandrou, S. (2014) *O Archaggelos tou Souniou*. Lemesos: Ieros Naos Archaggelou Michael.

Andriotis, K. (2009) Sacred site experience: A phenomenological study. *Annals of Tourism Research* 36 (1), 64–84.

Animals Australia (2019) *Go Forth and Travel with Kindness!* See https://www.animals australia.org/features/exploring-the-kind-way.php (accessed August 2019).

Apinunmahakul, A. and Devlin, R.A. (2008) Social networks and private philanthropy. *Journal of Public Economics* 92 (1–2), 309–328.

Ashworth, G.J. (2008) The memorialisation of violence and tragedy: Human trauma as heritage. In B. Graham and P. Howard (eds) *The Ashgate Research Companion to Heritage and Identity* (pp. 231–244). Aldershot: Ashgate.

Ashworth, G. and Hartmann, R. (2005) Introduction: Managing atrocity for tourism. In G. Ashworth and R. Hartmann (eds) *Horror and Human Tragedy Revisited: The Management of Sites of Atrocities for Tourism* (pp. 1–14). New York: Cognizant Communication Corporation.

Aslanidis, D. and Grigoriatis, D. (2001) *Apostle to Zaire: The Life and Legacy of Blessed Comas of Grigoriou*. Thessaloniki: Uncut Mountain Press.

Athletis (2018) *The Monastic Rule of St Basil the Great: Translated from the Original Greek*. Athletis Publishing (Kindle edn).

Auschwitz-Birkenau (2019) *Volunteers*. See http://auschwitz.org/en/volunteers/ (accessed November 2019).

Austin, A. (2016) On well-being and public policy: Are we capable of questioning the hegemony of happiness? *Social Indicators Research* 127 (1), 123–138.

Aydinoğlu, N.Z. and Krishna, A. (2010) Guiltless gluttony: The asymmetric effect of size labels on size perceptions and consumption. *Journal of Consumer Research* 37 (6), 1095–1112.

Bacon, F. (1985) *The Essays*. (J. Pitcher, ed.). Harmondsworth: Penguin.

Bakhtin, M. (1984) *Rabelais and his World* (H. Iswolsky, trans.). Bloomington, IN: Indiana University Press.

Bandyopadhyay, R. (2013) A paradigm shift in sex tourism research. *Tourism Management Perspectives* 6, 1–2.

Bandyopadhyay, R. (2018) Volunteer tourism and religion: The cult of Mother Teresa. *Annals of Tourism Research* 70, 133–136.

Barbieri, C., Santos, C.A. and Katsube, Y. (2012) Volunteer tourism: On-the-ground observations from Rwanda. *Tourism Management* 33 (3), 509–516.

Barnett, M.R. (1982) Nostalgia as nightmare: Blacks and American popular culture. *Crisis* 89 (2), 42–45.

Barney, R. (2010) Plato on the Desire for the Good. In S. Tenenbaum (ed.) *Desire, Practical Reason, and the Good* (pp. 34–64). Oxford: Oxford University Press.

Bartels, M. (2015) Genetics of wellbeing and its components satisfaction with life, happiness, and quality of life: A review and meta-analysis of heritability studies. *Behavior Genetics* 45 (2), 137–156.

Bartiromo, M. (2019) Anti-tourism group in Mallorca vandalizes rental cars to protest 'mass tourism'. *New York Post*, 7 August. See https://nypost.com/2019/08/07/anti-tourism-group-in-mallorca-vandalizes-rental-cars-to-protest-mass-tourism/ (accessed August 2019).

Bates, S. (2012) Cyprus: A true experience in north and south. *The Telegraph*, 1 April. See https://www.telegraph.co.uk/travel/destinations/europe/cyprus/articles0/Cyprus-a-true-experience-in-north-and-south/ (accessed October 2019).

Battour, M. and Ismail, M.N. (2016) Halal tourism: Concepts, practices, challenges and future. *Tourism Management Perspectives* 19, 150–154.

BBC (2006) Winter Solstice. *BBC Online*, 7 June. See https://www.bbc.co.uk/religion/religions/paganism/holydays/wintersolstice.shtml (accessed September 2019).

BBC (2012) Australian mourns 'angel' who saved 160 from suicide. *BBC News*, 15 May. See https://www.bbc.com/news/world-asia-18070939 (accessed November 2019).

BBC (2015) Denmark passes law to ban bestiality. *BBC Newsbeat*, 22 April. See http://www.bbc.co.uk/newsbeat/article/32411241/denmark-passes-law-to-ban-bestiality (accessed August 2019).

BBC (2016a) Japan region revises 'patronising' Chinese tourist guide. *BBC News*, 26 April. See https://www.bbc.com/news/blogs-news-from-elsewhere-36139230 (accessed September 2019).

BBC (2016b) Sex with animals remains banned in Germany as legal bid fails. *BBC News*, 19 February. See https://www.bbc.com/news/world-europe-35611906 (accessed August 2019).

BBC (2019) Uluru climbing ban: Tourists scale sacred rock for final time. *BBC News*, 25 October. See https://www.bbc.com/news/world-australia-50151344 (accessed November 2019).

BBC (2020) Coronavirus: Tourist hotspots deserted as virus spreads. *BBC News*, 6 March. See https://www.bbc.com/news/in-pictures-51768064 (accessed March 2020).

Beal, D., Trougakos, J., Weiss, H. and Green, S. (2006) Episodic processes in emotional labour: Perceptions of affective delivery and regulation strategies. *Journal of Applied Psychology* 91 (5), 1053–1065.

Beattie, J.M. and Schneider, I.E. (2018) Does service type influence satisfaction? A case study of Edinburgh Castle. *Tourism Management* 67, 89–97.

Bekkers, R. (2006) Traditional and health-related philanthropy: The role of resources and personality. *Social Psychology Quarterly* 69 (4), 349–366.

Bekkers, R. and Wiepking, P. (2011) A literature review of empirical studies of philanthropy: Eight mechanisms that drive charitable giving. *Nonprofit and Voluntary Sector Quarterly* 40 (5), 924–973.

Belfast Telegraph (2019) Northern Ireland scammers taking advantage of Thomas Cook collapse. *Belfast Telegraph Digital*, 27 September. See https://www.belfasttelegraph.co.uk/news/northern-ireland/northern-ireland-scammers-taking-advantage-of-thomas-cook-collapse-38539788.html (accessed September 2019).

Belhassen, Y. and Ebel, J. (2009) Tourism, faith and politics in the Holy Land: An ideological analysis of evangelical pilgrimage. *Current Issues in Tourism* 12 (4), 359–378.

Belhassen, Y., Caton, K. and Stewart, W.P. (2008) The search for authenticity in the pilgrim experience. *Annals of Tourism Research* 35 (3), 668–689.

Belhassen, Y., Cheer, J.M. and Kujawa, J. (2018) Why we still go on pilgrimages. *Journal of Tourism and Hospitality* 7 (6), 392–393.

Bell, B. (2014) Trek or treat: Should I pay for my friend's charity holiday? *BBC News*, 26 May. See https://www.bbc.com/news/uk-england-27485754 (accessed August 2019).

Bennett, M., King, B. and Milner, L. (2004) The health resort sector in Australia: A positioning study. *Journal of Vacation Marketing* 10 (2), 122–137.

Benson, J. (2013) *Environmental Ethics: An Introduction with Readings*. New York: Routledge.

Bentham, J. (1996 [1789]) An introduction to the principles of morals and legislation. In J. Burns and H.L.A. Hart (eds) *The Collected Works of Jeremy Bentham*. Oxford: Clarendon Press.

Bernstein, J.D. and Woosnam, K.M. (2019) Same same but different: Distinguishing what it means to teach English as a foreign language within the context of volunteer tourism. *Tourism Management* 72, 427–436.

Best, M. (2007) Norfolk Island: Thanatourism, history and visitor emotions. *Shima: The International Journal of Research into Island Cultures* 1 (2), 30–48.

Bialas, W. (2013) Nazi ethics: Perpetrators with a clear conscience. *Dapim: Studies on the Holocaust* 27 (1), 3–25.

Bigne, J.E., Andreu, L. and Gnoth, J. (2005) The theme park experience: An analysis of pleasure, arousal and satisfaction. *Tourism Management* 26 (6), 833–844.

Bimonte, S. and Faralla, V. (2012) Tourist types and happiness: A comparative study in Maremma, Italy. *Annals of Tourism Research* 39 (4), 1929–1950.

Bingaman, B. and Nassif, B. (eds) (2012) *The Philokalia: A Classic Text of Orthodox Spirituality*. Oxford: Oxford University Press.

Biran, A. and Hyde, K.F. (2013) New perspectives on dark tourism. *International Journal of Culture, Tourism and Hospitality Research* 7 (3), 191–198.

Blackwell, R. (2007) Motivations for religious tourism, pilgrimage, festivals and events. In R. Raj and N.D. Morpeth (eds) *Religious Tourism and Pilgrimage Festivals Management: An International Perspective* (pp. 35–47). Wallingford: CABI.

Blain, M. and Lashley, C. (2014) Hospitableness: The new service metaphor? Developing an instrument for measuring hosting. *Research in Hospitality Management* 4 (1–2), 1–8.

Bloom, W. (1990) *Personal Identity, National Identity and International Relations*. Cambridge: Cambridge University Press.

Boehm, J.K. and Lyubomirsky, S. (2008) Does happiness promote career success? *Journal of Career Assessment* 16 (1), 101–116.

Bojanowska, A. and Zalewska, A.M. (2016) Lay understanding of happiness and the experience of well-being: Are some conceptions of happiness more beneficial than others? *Journal of Happiness Studies* 17 (2), 793–815.

Bond, N., Packer, J. and Ballantyne, R. (2015) Exploring visitor experiences, activities and benefits at three religious tourism sites. *International Journal of Tourism Research* 17 (5), 471–481.

Boniwell, I. (2008) *Positive Psychology in a Nutshell: A Balanced Introduction to the Science of Optimal Functioning* (2nd edn). London: Personal Well-Being Centre.

Borowitz, E.B. and Schwartz, F.W. (1999) *The Jewish Moral Virtues*. Philadelphia, PA: Jewish Publication Society.

Botsari, D. (2014) *Pemptousia*. See https://pemptousia.com (accessed March 2019).

Bove, L.L. (2019) Empathy for service: Benefits, unintended consequences, and future research agenda. *Journal of Services Marketing* 33 (1), 31–43.

Bowman, M.S. and Pezzullo, P.C. (2010) What's so 'dark' about 'dark tourism'?: Death, tours, and performance. *Tourist Studies* 9 (3), 187–202.

Boyles, F. (2005) Andersonville: A site steeped in controversy. In G. Ashworth and R. Hartmann (eds) *Horror and Human Tragedy Revisited: The Management of Sites of Atrocities for Tourism* (pp. 73–85). New York: Cognizant Communication Corporation.

Brammer, S. and Millington, A. (2005) Corporate reputation and philanthropy: An empirical analysis. *Journal of Business Ethics* 61 (1), 29–44.

Branigan, T. (2013) Chinese tourists warned over bad behavior overseas. *The Guardian*, 17 May. See https://www.theguardian.com/world/2013/may/17/chinese-tourists-warned-behaving-badly-wang-yang (accessed September 2019).

Brooker, P. (1987) Review. Paul Spoonley, The Politics of Nostalgia: Racism and the Extreme Right in New Zealand (Palmerston North: Dunmore Press, 1987). *Political Science* 39 (2), 194–196.

Brotheridge, C.M. and Lee, R.T. (2008) The emotions of managing: An introduction to the special issue. *Journal of Managerial Psychology* 23 (2), 108–117.

Brotherton, B. (2005) The nature of hospitality: Customer perceptions and implications. *Tourism and Hospitality Planning & Development* 2 (3), 139–153.

Brownless, J. (2019) The Fab 50 – Ireland's 50 best places to stay in 2019. *Independent.ie*. See https://www.independent.ie/editorial/StoryPlus/indofab50/ (accessed August 2019).

Brummer, A. (2017) How my blood boils when grasping airlines treat us all like cash cows: City editor Alex Brummer says budget flight carriers have given up on pretence of politeness. *Daily Mail*, 19 September. See https://www.dailymail.co.uk/debate/article-4897364/Easyjet-Budget-airlines-given-politeness-pretence.html (accessed August 2019).

Bruner, E.M. (1991) Transformation of self in tourism. *Annals of Tourism Research* 18 (2), 238–250.

Buda, D.M. (2015) The death drive in tourism studies. *Annals of Tourism Research* 50, 39–51.

Buda, D.M., d'Hauteserre, A. and Johnston, L. (2014) Feeling and tourism studies. *Annals of Tourism Research* 46, 102–114.

Budiani-Saberi, D.A. and Delmonico, F.L. (2008) Organ trafficking and transplant tourism: A commentary on the global realities. *American Journal of Transplantation* 8 (5), 925–929.

Buntman, B. (2008) Tourism and tragedy: The memorial at Belzec, Poland. *International Journal of Heritage Studies* 14 (5), 422–448.

Burgen, S. (2018) How tourism is killing Barcelona. *The Guardian*, 30 August. See https://www.theguardian.com/travel/2018/aug/30/why-tourism-is-killing-barcelona-over-tourism-photo-essay (accessed September 2019).

Burns, G.L. (2015) Ethics in tourism. In C.M. Hall, S. Gössling and D. Scott (eds) *The Routledge Handbook of Tourism and Sustainability* (pp. 117–126). London: Routledge.

Butcher, J. (2005) *The Moralisation of Tourism: Sun, Sand ... and Saving the World?* London and New York: Routledge.

Butcher, J. (2015) Ethical tourism and development: The personal and the political. *Tourism Recreation Research* 40 (1), 71–80.

Butler, R. (1980) The concept of a tourist area cycle of evolution: Implications for management of resources. *Canadian Geographer* 24, 5–12.

Butler, R. (2009) Tourism in the future: Cycles, waves or wheels? *Futures* 41, 346–352.

Cabezas, A.L. (2004) Between love and money: Sex, tourism, and citizenship in Cuba and the Dominican Republic. *Signs* 29 (4), 987–1015.

Cafaro, P. (2005) Gluttony, arrogance, greed, and apathy: An exploration of environmental vice. In P. Cafaro and R. Sandler (eds) *Environmental Virtue Ethics* (pp. 135–158). Lanham, MD: Rowman & Littlefield.

Calder, S. (2015) How to tip in restaurants: In some countries tipping is simply not done. *Independent*, 20 November. See https://www.independent.co.uk/travel/news-and-advice/how-to-tip-in-restaurants-in-some-countries-tipping-is-simply-not-done-a6741616.html (accessed August 2019).

Campelo, A., Aitken, R. and Gnoth, J. (2011) Visual rhetoric and ethics in marketing of destinations. *Journal of Travel Research* 50 (1), 3–14.

Campelo, A., Aitken, R., Thyne, M. and Gnoth, J. (2014) Sense of place: The importance for destination branding. *Journal of Travel Research* 53, 154–166.

Cannon, W.B. (1927) The James–Lange theory of emotion: A critical examination and an alternative theory. *American Journal of Psychology* 39, 106–124.

Canosa, A. and Graham, A. (2016) Ethical tourism research involving children. *Annals of Tourism Research* 61 (15), 1–6.

Carlisle, S., Henderson, G. and Hanlon, P.W. (2009) Wellbeing: A collateral casualty of modernity? *Social Science and Medicine* 69, 1556–1560.

Carneiro, M.J., Lima, J. and Silva, A.L. (2015) Landscape and the rural tourism experience: Identifying key elements, addressing potential, and implications for the future. *Journal of Sustainable Tourism* 23 (8–9), 1217–1235.

Caroll, A.B (1991) The pyramid of corporate social responsibility: Toward the moral management of organizational stakeholders. *Business Horizons* 34 (4), 39–48.

Cater, C.I. (2006) Playing with risk? Participant perceptions of risk and management implications in adventure tourism. *Tourism Management* 27 (2), 317–325.

Caton, K. (2012) Taking the moral turn in tourism studies. *Annals of Tourism Research* 39 (4), 1906–1928.

Caton, K., Mair, H., Muldoon, N. and Grimwood, B.S.R. (2018) Introduction: Wellbeing conceptualized across the disciplines. In B.S.R. Grimwood, H. Mair, K. Caton and M. Muldoon (eds) (2018) *Tourism and Wellness: Travel for the Good of All?* Lanham, MD: Rowman & Littlefield.

Cavarnos, C. (1994) *St. Nicodemos the Hagiorite*. Belmont, MA: Institute for Byzantine and Modern Greek Studies.

Cetin, G. and Okumus, F. (2018) Experiencing local Turkish hospitality in Istanbul, Turkey. *International Journal of Culture, Tourism and Hospitality Research* 12 (2), 223–237.

Chakrabortty, A. (2019) The lesson from the ruins of Notre Dame: Don't rely on billionaires. *The Guardian*, 18 July. See https://www.theguardian.com/commentisfree/2019/jul/18/ruins-notre-dame-billionaires-french-philanthropy (accessed August 2019).

Chapin, F.S. and Knapp, C.N. (2015) Sense of place: A process for identifying and negotiating potentially contested visions of sustainability. *Environmental Science and Policy* 53, 38–46.

Chapman, A. and Light, D. (2017) Working with the carnivalesque at the seaside: Transgression and misbehaviour in a tourism workplace. *Tourist Studies* 17 (2), 182–199.

Cheer, J.M., Belhassen, Y. and Kujawa, J.M. (2017) Spiritual tourism: Entrée to the special issue. *Tourism Management Perspectives* 24, 186–187.

Chemin, M. and Mbiekop, F. (2015) Addressing child sex tourism: The Indian case. *European Journal of Political Economy* 38, 169–180.

Chen, A. and Peng, N. (2014) Examining Chinese consumers' luxury hotel staying behavior. *International Journal of Hospitality Management* 39, 53–56.

Chen, M. and Lin C. (2015) Understanding corporate philanthropy in the hospitality industry. *International Journal of Hospitality Management* 48, 150–160.

Chen, Y., Lehto, X.Y. and Cai, L. (2013) Vacation and well-being: A study of Chinese tourists. *Annals of Tourism Research* 42, 284–310.

Chessick, R.D. (1965) Empathy and love in psychotherapy. *American Journal of Psychotherapy* 19 (2), 205–219.

Cheung, W.Y., Sedikides, C. and Wildschut, T. (2016) Induced nostalgia increases optimism (via social-connectedness and self-esteem) among individuals high, but not low, in trait nostalgia. *Personality and Individual Differences* 90, 283–288.
Chondropoulos, S. (1997) *Saint Nektarios: The Saint of our Century*. Athens: Kainourgia Ge.
Chou, H. and Lien, N. (2010) Advertising effects of songs' nostalgia and lyrics' relevance. *Asia Pacific Journal of Marketing and Logistics* 22 (3), 314–329.
Christou, P. (2016) Transcending the limits of hospitality: The case of Mount Athos and the offering of philoxenia. In C. Lashley (ed.) *The Routledge Handbook of Hospitality Studies* (pp. 337–347). New York: Routledge.
Christou, P. (2018) Exploring *agape*: Tourists on the island of love. *Tourism Management* 68, 13–22.
Christou, P.A. (2019a) Einstein's theory of relativity informing research relating to social sciences, tourism and the tourist experience. *Current Issues in Tourism* 1–7.
Christou, P. (2019b) Neoteny: The paedomorphosis of destinations. *Annals of Tourism Research* 81 (C). doi:10.1016/j.annals.2019.03.004
Christou, P. (2020a) Tourism experiences as the remedy to nostalgia: Conceptualizing the nostalgia and tourism nexus. *Current Issues in Tourism* 23 (5), 612–625. doi:10.1080/13683500.2018.1548582
Christou, P. (2020b) Place disbelief: A tourism-experiential perspective. *Annals of Tourism Research* 80, 102676.
Christou, P.A. and Farmaki, A. (2019) Utopia as a reinforcement of tourist experiences. *Annals of Tourism Research* 77, 144–147. doi:10.1016/j.annals.2018.11.003
Christou, P. and Sharpley, R. (2019) *Philoxenia* offered to tourists? A rural tourism perspective. *Tourism Management* 72, 39–51.
Christou, P. and Simillidou, A. (2020) Tourist experience: The catalyst role of tourism in comforting melancholy, or not. *Journal of Hospitality and Tourism Management* 42, 210–221.
Christou, P., Farmaki, A. and Evangelou, G. (2018a) Nurturing nostalgia? A response from rural tourism stakeholders. *Tourism Management* 69, 42–51.
Christou, P., Sharpley, R. and Farmaki, A. (2018b) Exploring the emotional dimension of visitors' satisfaction at cultural events. *Event Management* 22 (2), 255–269.
Christou, P., Avloniti, A. and Farmaki, A. (2019a) Guests' perceptions of emotionally expressive and non-expressive service providers within the hospitality context. *International Journal of Hospitality Management* 76, 152–162.
Christou, P., Hadjielas, E. and Farmaki, A. (2019b) Reconnaissance of philanthropy. *Annals of Tourism Research* 78, 102749.
Christou, P.A., Farmaki, A., Saveriades, A. and Spanou, E. (2019c) The 'genius loci' of places that experience intense tourism development. *Tourism Management Perspectives* 30, 19–32.
Claviez, T. (ed.) (2013) *The Conditions of Hospitality: Ethics, Politics, and Aesthetics on the Threshold of the Possible*. New York: Fordham University Press.
Cloninger, R.C. (2004) *Feeling Good: The Science of Well-being*. Oxford: Oxford University Press.
CN Traveler (2015a) What not to do ... anywhere: 6 bad travel behaviors to avoid. *Condé Nast Traveler*, 9 March. See https://www.cntraveler.com/galleries/2015-03-09/what-not-to-do-anywhere-6-bad-travel-behaviors-to-avoid (accessed September 2019).
CN Traveler (2015b) 12 amazing things hotels have done for their guests. *Condé Nast Traveler*, 28 April. See https://www.cntraveller.in/story/12-amazing-things-hotels-have-done-their-guests/ (accessed December 2019).
Coffey, H. (2019) Rome bans tourists from going topless, eating 'messy' food and leaving love padlocks. *Independent*, 10 June. See https://www.independent.co.uk/travel/news-and-advice/rome-tourist-rules-italy-topless-ban-water-fountains-love-padlocks-fines-a8951766.html (accessed September 2019).

Coghlan, A. (2015a) Prosocial behaviour in volunteer tourism. *Annals of Tourism Research* 55, 46–60.
Coghlan, A. (2015b) Tourism and health: Using positive psychology principles to maximise participants' wellbeing outcomes – a design concept for charity challenge tourism. *Journal of Sustainable Tourism* 23 (3), 382–400.
Cohen, E. (1974) Who is a tourist? A conceptual clarification. *The Sociological Review* 22, 533 and 547.
Cohen, E. (1979) A phenomenology of tourist experiences. *Sociology* 13, 179–201.
Cohen, E. (1988) Authenticity and commoditization in tourism. *Annals of Tourism Research* 15 (3), 371–386.
Cohen, E.H. (2011) Educational dark tourism at an 'in populo' site: The Holocaust Museum in Jerusalem. *Annals of Tourism Research* 38 (1), 193–209.
Cohen, E. (2018) Thanatourism: A comparative approach. In P.R. Stone, R. Hartmann, T. Seaton, R. Sharpley and L. White (eds) *The Palgrave Handbook of Dark Tourism Studies* (pp. 157–171). London: Palgrave Macmillan.
Cohen, S.A. and Cohen, E. (2019) New directions in the sociology of tourism. *Current Issues in Tourism* 22 (2), 153–172.
Colau, A. (2014) Mass tourism can kill a city – just ask Barcelona's residents. *The Guardian*, 2 September. See https://www.theguardian.com/commentisfree/2014/sep/02/mass-tourism-kill-city-barcelona (accessed November 2019).
Coldwell, W. (2017) First Venice and Barcelona: Now anti-tourism marches spread across Europe. *The Guardian*, 10 August. See https://www.theguardian.com/travel/2017/aug/10/anti-tourism-marches-spread-across-europe-venice-barcelona (accessed November 2019).
Collins-Kreiner, N., Kliot, N., Mansfeld, Y. and Sagi, K. (2017) *Christian Tourism to the Holy Land: Pilgrimage During Security Crisis*. New York: Routledge.
Compton, N.B. (2019) A selfie ban in the Czech Republic is the latest effort to combat bad tourist behavior. *Washington Post*, 18 October. See https://www.washingtonpost.com/travel/2019/10/18/selfie-ban-czech-republic-is-latest-effort-combat-bad-tourist-behavior/ (accessed November 2019).
Constantelos, D.J. (1968) *Byzantine Philanthropy and Social Welfare*. New Brunswick, NJ: Rutgers University Press.
Constantelos, D.J. (2016) Byzantine philanthropy. *Pemptousia*, 25 August. See http://pemptousia.com/2016/08/byzantine-philanthropy-part-i/ (accessed March 2018).
Cook, C.C.H. (2010) The Philokalia and mental wellbeing. Doctoral dissertation, Durham University.
Cook, W. (2012) For men only: A pilgrimage to Mount Athos in Greece. *Independent*, 26 May. See https://www.independent.co.uk/travel/europe/for-men-only-a-pilgrimage-to-mount-athos-in-greece-7786356.html (accessed October 2019).
Cooke, P.J., Melchert, T.P. and Connor, K. (2016) Measuring well-being: A review of instruments. *The Counseling Psychologist* 44 (5), 730–757.
Costa, G., Glinia, E. and Drakou, A. (2004) The role of empathy in sport tourism services: A review. *Journal of Sport & Tourism* 9 (4), 331–342.
Coudounaris, D.N. and Sthapit, E. (2017) Antecedents of memorable tourism experience related to behavioral intentions. *Psychology and Marketing* 34 (12), 1084–1093.
Creswell, T. (2004) *Place: A Short Introduction*. Oxford: Blackwell.
Crompton, J.I. (1979) Motivations for pleasure vacation. *Annals of Tourism Research* 6 (4), 408–424.
Cuffy, V. (2017) Carnival tourism. In S. Agarwal, G. Busby and R. Huang (eds) *Special Interest Tourism: Concepts, Contexts and Cases* (pp. 97–111). Wallingford: CABI.
Dale, C. and Robinson, N. (2011) Dark tourism. In P. Robinson, S. Heitmann and P.U.C. Dieke (eds) *Research Themes for Tourism* (pp. 205–217). Wallingford: CABI.

Dann, G.M. (1981) Tourist motivation: An appraisal. *Annals of Tourism Research* 8 (2), 187–219.
Dann, G.M. (1995) Tourism: The nostalgia industry of the future. In W.F. Theobald (ed.) *Global Tourism: The Next Decade* (pp. 55–67). Oxford: Butterworth-Heinemann.
Dann, G. (1996) *The Language of Tourism: A Sociolinguistic Perspective*. Wallingford: CABI.
Davalos, S., Merchant, A., Rose, G.M., Lessley, B.J. and Teredesai, A.M. (2015) The good old days: An examination of nostalgia in Facebook posts. *International Journal of Human-Computer Studies* 83, 83–93.
Davis, N. (2019) Euthanasia and assisted dying rates are soaring. But where are they legal? *The Guardian*, 15 July. See https://www.theguardian.com/news/2019/jul/15/euthanasia-and-assisted-dying-rates-are-soaring-but-where-are-they-legal (accessed November 2019).
Day, C. (2019) 10 weird rules Disney princesses need to follow at the parks. *The Travel*, 27 August. See https://www.thetravel.com/weird-rules-disney-princesses-follow-parks/ (accessed August 2019).
De Bloom, J., Geurts, S.A., Taris, T.W., Sonnentag, S., de Weerth, C. and Kompier, M.A. (2010) Effects of vacation from work on health and well-being: Lots of fun, quickly gone. *Work and Stress* 24 (2), 196–216.
De Groot, J.H., Smeets, M.A., Rowson, M.J., Bulsing, P.J., Blonk, C.G., Wilkinson, J.E. and Semin, G.R. (2015) A sniff of happiness. *Psychological Science* 26 (6), 684–700.
De Jong, R. (2018) Sip an iconic Singapore Sling where it all started. *Lonely Planet*, 31 October. See https://www.lonelyplanet.com/news/2018/10/31/raffles-hotels-singapore-sling-long-bar/ (accessed August 2019).
De Jong, A. and Varley, P. (2017) Food tourism policy: Deconstructing boundaries of taste and class. *Tourism Management* 60, 212–222.
Dearden, L. (2014) Qatar launches campaign for 'modest' dress code for tourists. *Independent*, 27 May. See https://www.independent.co.uk/news/world/middle-east/qatar-launches-campaign-for-modest-dress-code-for-tourists-9438452.html (accessed November 2019).
Dekker, D.M. (2014) Personality and hospitable behavior. In I. Pantelides (ed.) *The Routledge Handbook of Hospitality Management* (pp. 57–66). Abingdon: Routledge.
Della Dora, V. (2012) Setting and blurring boundaries: Pilgrims, tourists, and landscape in Mount Athos and Meteora. *Annals of Tourism Research* 39 (2), 951–974.
Demetriou, L. (2012) From philoxenia to xenophobia? Relations between xenophobic tendencies and parental acceptance-rejection childhood experiences for Greek Cypriot university students. *International Journal of Social Sciences and Education* 3 (2), 296–316.
Der Spiegel (2018) Paradise lost: How tourists are destroying the places they love. *Der Spiegel*, 21 August. See https://www.spiegel.de/international/paradise-lost-tourists-are-destroying-the-places-they-love-a-1223502.html (accessed September 2019).
Derrida, J. (2002) *Acts of Religion*. London: Routledge.
Deutsche Welle (2017a) Selfie with Hitler: Indonesia wax museum removes Nazi-themed exhibit. *Deutsche Welle*, 11 November. See https://www.dw.com/en/selfie-with-hitler-indonesia-wax-museum-removes-nazi-themed-exhibit/a-41342077 (accessed November 2019).
Deutsche Welle (2017b) Selfies at Dachau: New film reveals embarrassing reality of remembrance. *Deutsche Welle*, 27 January. See https://www.dw.com/en/selfies-at-dachau-new-film-reveals-embarrassing-reality-of-remembrance/a-37295443 (accessed November 2019).
Dhar, R.L. (2015) Service quality and the training of employees: The mediating role of organizational commitment. *Tourism Management* 46, 419–430.

Diamond, M. and Olito, F. (2019) 17 places around the world that are being ruined by tourism. *Insider*, 28 October. See https://www.insider.com/cities-hurt-by-tourism-2017-12 (accessed November 2019).

Dickinson, G. (2018) 'Unlimited, disrespectful and excessive' – is the party over for tourism in Ibiza? *The Telegraph*, 24 April. See https://www.telegraph.co.uk/travel/news/ibiza-tourism-protest/ (accessed September 2019).

Dickinson, G. (2019) Japanese attractions ban foreigners in overtourism backlash. *The Telegraph*, 7 March. See https://www.telegraph.co.uk/travel/news/japan-overtourism-banning-tourists/ (accessed September 2019).

Diekmann, A. and Hannam, K. (2012) Touristic mobilities in India's slum spaces. *Annals of Tourism Research* 39 (3), 1315–1336.

Diener, E. and Seligman, M.E.P. (2002) Very happy people. *Psychological Science* 13, 81–84.

Diener, E., Lucas, R.E. and Oishi, S. (2002) Subjective well-being: The science of happiness and life satisfaction. In S.J. Lopez and C.R. Snyder (eds) *Handbook of Positive Psychology* (pp. 63–73). Oxford: Oxford University Press.

Diener, E., Oishi, S. and Tay, L. (2018) Advances in subjective well-being research. *Nature Human Behaviour* 2 (4), 253.

Dimitriou, C.K. (2017) The quest for a practical approach to morality and the tourism industry. *Journal of Hospitality and Tourism Management* 31, 45–51.

Dimitrovski, D. and Todorović, A. (2015) Clustering wellness tourists in spa environment. *Tourism Management Perspectives* 16, 259–265.

Dinhopl, A. and Gretzel, U. (2016) Selfie-taking as touristic looking. *Annals of Tourism Research* 57, 126–139.

Dolan, R., Seo, Y. and Kemper, J. (2019) Complaining practices on social media in tourism: A value co-creation and co-destruction perspective. *Tourism Management* 73, 35–45.

Dorothy, J. (2016) The elephant who helped me through one of the darkest periods of my life. *The Guardian*, 31 October. See https://www.theguardian.com/environment/2016/oct/31/traumatised-elephant-helped-me-overcome-abuse (accessed August 2019).

Dossey, L. (2010) Gluttony and obesity – Editorial. *EXPLORE: The Journal of Science and Healing* 6 (1), 1–6.

Doxey, G. (1975) A causation theory of visitor–resident irritants: Methodology and research inferences. The impact of tourism. In *Sixth Annual Conference Proceedings of the Travel Research Association* (pp. 195–198). San Diego, CA: Travel Research Association.

Drennan, J., Bianchi, C., Cacho-Elizondo, S., Louriero, S., Guibert, N. and Proud, W. (2015) Examining the role of wine brand love on brand loyalty: A multi-country comparison. *International Journal of Hospitality Management* 49, 47–55.

Dunkley, R.A., Morgan, N. and Westwood, S. (2007) A shot in the dark? Developing a new conceptual framework for thanatourism. *Asian Journal of Tourism and Hospitality* 1 (1), 54–63.

Durant, W. (2012) *The adventure of philosophy*. Metaichmio publications.

Edensor, T. (2007) Mundane mobilities, performances and spaces of tourism. *Social and Cultural Geography* 8 (2), 199–215.

Edensor, T. (2018) The more-than-visual experiences of tourism. *Tourism Geographies* 20 (5), 913–915.

Edwards, F. and Mercer, D. (2007) Gleaning from gluttony: An Australian youth subculture confronts the ethics of waste. *Australian Geographer* 38 (3), 279–296.

Ekathimerini (2018) Years later, remorseful tourist returns stones taken from Acropolis site. *Ekathimerini*, 5 February. See http://www.ekathimerini.com/225514/article/ekathimerini/news/years-later-remorseful-tourist-returns-stones-taken-from-acropolis-site (accessed October 2019).

Ekman, P. (ed.) (1973) *Darwin and Facial Expression*. New York: Academic Press.

Ekman, P. (1982) *Emotion in the Human Face* (2nd edn). Cambridge: Cambridge University Press.

Ekman, P. (2016) What scientists who study emotion agree about. *Perspectives on Psychological Science* 11 (1), 31–34.

Ekman, P. and Cordaro, D. (2011) What is meant by calling emotions basic. *Emotion Review* 3 (4), 364–370.

Ekman, P., Friesen, W. and Ellsworth, P. (1972) *Emotion in the Human Face: Guidelines for Research and an Integration of Findings*. New York: Pergamon Press.

Elbaz, A.M. and Haddoud, M.Y. (2017) The role of wisdom leadership in increasing job performance: Evidence from the Egyptian tourism sector. *Tourism Management* 63, 66–76.

Ellis, A., Park, E., Kim, S. and Yeoman, I. (2018) What is food tourism? *Tourism Management* 68, 250–263.

Emirates (2019a) *Emirates Service Reviews*. See https://www.emirates.com/cy/english/experience/review/service/ (accessed August 2019).

Emirates (2019b) *The Emirates Airline Foundation*. See http://www.emiratesairlinefoundation.org (accessed August 2019).

Enghagen, L.K. (1990) Teaching ethics in hospitality and tourism education. *Hospitality Research Journal* 14 (2), 467–474.

Etkin, J. and Mogilner, C. (2016) Does variety among activities increase happiness? *Journal of Consumer Research* 43 (2), 210–229.

Euronews (2019) Philadelphia, Pennsylvania: The birthplace of US democracy … and the legendary cheesesteak. *Euronews*, 18 March. See https://www.euronews.com/2019/03/18/philadelphia-pennsylvania-the-birthplace-of-us-democracy-and-the-legendary-cheesesteak (accessed August 2019).

Evripidou, N. (2010) The 'marriage' of carnival with Christianity. *Sigmalive*, 7 February. See (in Greek) https://www.sigmalive.com/archive/simerini/news/social/235623 (accessed September 2019).

Explore (2020) Review: Real journeys' magical doubtful sound overnight cruise. *Explore Now or Never*, 15 May. See https://explorenowornever.com/doubtful-sound-overnight-cruise/.

Fantozzi, J. (2018) 22 secrets every Disney World lover should know. *Insider*, 11 September. See https://www.insider.com/disney-world-secrets-facts-2018-1 (accessed August 2019).

Farasiotis, D. (2005) *The Gurus, the Young and Elder Paisios* (7th edn). Thessaloniki: St. Herman of Alaska Brotherhood.

Farmaki, A. (2018) Corporate social responsibility in hotels: A stakeholder approach. *International Journal of Contemporary Hospitality Management* 31 (6), 2297–2320.

Farmaki, A. and Antoniou, K. (2017) Politicising dark tourism sites: Evidence from Cyprus. *Worldwide Hospitality and Tourism Themes* 9 (2), 175–186.

Farmaki, A. and Farmakis, P. (2018) A stakeholder approach to CSR in hotels. *Annals of Tourism Research* 68, 58–60.

Farmaki, A., Georgiou, M. and Christou, P. (2017) Growth and impacts of all-inclusive holiday packages: Echoes from the industry. *Tourism Planning and Development* 14 (4), 483–502.

Farmaki, A., Khalilzadeh, J. and Altinay, L. (2019) Travel motivation and demotivation within politically unstable nations. *Tourism Management Perspectives* 29, 118–130.

Faullant, R., Matzler, K. and Mooradian, T.A. (2011) Personality, basic emotions, and satisfaction: Primary emotions in the mountaineering experience. *Tourism Management* 32 (6), 1423–1430.

Fennell, D.A. (2006) Evolution in tourism: The theory of reciprocal altruism and tourist–host interactions. *Current Issues in Tourism* 9 (2),105–124.

Fennell, D.A. (2011) *Tourism and Animal Ethics*. New York: Routledge.

Fennell, D.A. (2014) Exploring the boundaries of a new moral order for tourism's global code of ethics: An opinion piece on the position of animals in the tourism industry. *Journal of Sustainable Tourism* 22 (7), 983–996.

Fennell, D.A. (2015) Ethics in tourism. In G. Moscardo and P. Benckendorff (eds) *Education for Sustainability in Tourism* (pp. 45–57). Berlin and Heidelberg: Springer.

Fennell, D.A. and Malloy, D.C. (1999) Measuring the ethical nature of tourism operators. *Annals of Tourism Research* 26 (4), 928–943.

Fennell, D.A. and Malloy, D. (2007) *Codes of Ethics in Tourism: Practice, Theory, Synthesis*. Clevedon: Channel View Publications.

Feodorovna, E. (2005) *The Martha-Mary Convent: and Rule of St. Elizabeth the New Martyr*. The Printshop of St Job of Pochaev. New York: Holy Trinity Monastery.

Ferguson, G., Megehee, C.M. and Woodside, A.G. (2017) Culture, religiosity, and economic configural models explaining tipping-behavior prevalence across nations. *Tourism Management* 62, 218–233.

Fetscherin, M. and Stephano, R.M. (2016) The medical tourism index: Scale development and validation. *Tourism Management* 52, 539–556.

Filep, S. (2008) Measuring happiness: A new look at tourist satisfaction. In S. Richardson, L. Fredline, A. Patiar and M. Ternel (eds) *CAUTHE 2008: Tourism and Hospitality Research, Training and Practice; 'Where the Bloody Hell Are We?'* (pp. 13–19). Gold Coast: Griffith University.

Filimonau, V. and De Coteau, D.A. (2019) Food waste management in hospitality operations: A critical review. *Tourism Management* 71, 234–246.

Fiorini, F. and Amiel, S. (2019) Watch: Bologna's homeless show tourists a different side of the city. *Euronews*, 30 July. See https://www.euronews.com/2019/07/30/watch-bologna-s-homeless-show-tourists-a-different-side-to-the-city (accessed August 2019).

Fleischer, A. and Rivlin, J. (2008) More or better?: Quantity and quality issues in tourism consumption. *Journal of Travel Research* 47 (3), 285–294.

Flesoras, C. (2011) Dressing the soul and the body. *Metropolis of San Francisco Shrine*, 9 January. See https://www.saintanna.org/kairos/2016/4/13/dressing-the-soul-and-the-body (accessed November 2019).

Ford, B.Q., Dmitrieva, J.O., Heller, D., *et al.* (2015) Culture shapes whether the pursuit of happiness predicts higher or lower well-being. *Journal of Experimental Psychology: General* 144 (6), 1053–1062.

Forest of Bowland (2015) *Sense of Place*. See https://www.forestofbowland.com/Sense-Place (accessed September 2019).

Forest of Bowland (2019) *Sense of Place Toolkit*. See https://www.forestofbowland.com/files/uploads/pdfs/sense_of_place_final2.pdf.

Fox, K. (2019) Venice becomes the front line in the battle against overtourism. *CNN Travel*, 15 June. See https://edition.cnn.com/travel/article/venice-tourism-overcrowding-intl/index.html (accessed August 2019).

Fredrickson, B. (2001) The role of positive emotions in positive psychology: The broaden-and-build theory of positive emotions. *American Psychologist* 56 (12), 218–226.

Frenzel, F., Koens, K. and Steinbrink, M. (eds) (2012) *Slum Tourism: Poverty, Power and Ethics*. New York: Routledge.

Freud, S. (1984) *On Metapsychology, the Theory of Psychoanalysis: Beyond the Pleasure Principle, the Ego and the Id, and Other Works, Vol. 2* (J. Strachey, trans.). Harmondsworth: Penguin.

Friedrich, M. and Johnston, T. (2013) Beauty versus tragedy: Thanatourism and the memorialisation of the 1994 Rwandan Genocide. *Journal of Tourism and Cultural Change* 11 (4), 302–320.

Fromm, E. (2008) *The Art of Loving: The Centennial Edition*. New York: Continuum International.
Gao, J. and Kerstetter, D.L. (2018) From sad to happy to happier: Emotion regulation strategies used during a vacation. *Annals of Tourism Research* 69, 1–14.
Garcia, B.R. (2012) Management issues in dark tourism attractions: The case of ghost tours in Edinburgh and Toledo. *Journal of Unconventional Parks, Tourism and Recreation Research* 4 (1), 14–19.
Gardiner, S. and Kwek, A. (2017) Chinese participation in adventure tourism: A study of generation Y international students' perceptions. *Journal of Travel Research* 56 (4), 496–506.
Gaur, S.S., Herjanto, H. and Makkar, M. (2014) Review of emotions research in marketing, 2002–2013. *Journal of Retailing and Consumer Services* 21 (6), 917–923.
Gauthier, S., Mausbach, J., Reisch, T. and Bartsch, C. (2015) Suicide tourism: A pilot study on the Swiss phenomenon. *Journal of Medical Ethics* 41 (8), 611–617.
Gayley, C.M. (1893) *The Classic Myths in English Literature and in Art*. Boston, MA: Ginn.
Gehrels, S. (2017) Liquid hospitality: Wine as the metaphor. In C. Lashley (ed.) *The Routledge Handbook of Hospitality Studies* (pp. 247–259). Abingdon: Routledge.
Genius Loci (2019) *Genius Loci – Co-founded European Project under the COSME Programme of the European Union*. See http://www.europeangeniusloci.eu/ (accessed September 2019).
Gentzler, A.L., Palmer, C.A., Ford, B.Q., Moran, K.M. and Mauss, I.B. (2019) Valuing happiness in youth: Associations with depressive symptoms and well-being. *Journal of Applied Developmental Psychology* 62, 220–230.
Gilbert, D. and Abdullah, J. (2004) Holidaytaking and the sense of well-being. *Annals of Tourism Research* 31 (1), 103–121.
Gill, C., Packer, J. and Ballantyne, R. (2019) Spiritual retreats as a restorative destination: Design factors facilitating restorative outcomes. *Annals of Tourism Research* 79, 102761.
Gillespie, T. (2016) Pigs might fly: Ryanair slammed for 'Fly to Win' charity scratch card where odds of winning are 1.2 billion/1 … and only small percentage of the profits go to good causes. *The Sun*, 15 October. See https://www.thesun.co.uk/news/1982591/ryanair-slammed-for-fly-to-win-charity-scratch-card-where-odds-of-winning-are-1-2billion1-and-only-small-percentage-of-the-profits-go-to-good-causes/ (accessed August 2019).
Gillet, S., Schmitz, P. and Mitas, O. (2016) The snap-happy tourist: The effects of photographing behavior on tourists' happiness. *Journal of Hospitality and Tourism Research* 40 (1), 37–57.
Gilovich, T., Kumar, A. and Jampol, L. (2015) A wonderful life: Experiential consumption and the pursuit of happiness. *Journal of Consumer Psychology* 25 (1), 152–165.
Glen, P. (2013) Is philanthropy the future of tourism? *Calgary Herald*, 1 May. See https://calgaryherald.com/travel/is-philanthropy-the-future-of-tourism (accessed August 2019).
Global Wellness Institute (2019) *Global Wellness Economy Monitor Press Release*. See https://globalwellnessinstitute.org/press-room/press-releases/global-wellness-institute-releases-global-wellness-economy-monitor-packed-with-regional-national-data-on-wellness-markets/ (accessed October 2019).
Godfrey, J. (1984) I love New York. *Tourism Management* 5 (2), 148–149.
Godfrey, K. (2019a) De-feeted: Passenger's disgust at man who had his bare feet on the in-flight entertainment screen for so long he left toe prints. *The Sun*, 20 August. See https://www.thesun.co.uk/travel/9758522/passenger-bare-feet-plane-screen/ (accessed August 2019).
Godfrey, K. (2019b) Sands nasty: Stunning beach closed after tourist spotted burying a nappy in the sand. *The Sun*, 15 August. See https://www.thesun.co.uk/travel/9726482/beach-closed-boracay-tourist-buries-nappy/ (accessed August 2019).

Godfrey, K. (2019c) 'Pure gold'. *The Sun*, 23 September. See https://www.thesun.co.uk/travel/9985142/thomas-cook-collapse-staff-work-unemployed/ (accessed September 2019).

Goldstein, D.M. and Hall, K. (2017) Postelection surrealism and nostalgic racism in the hands of Donald Trump. *HAU: Journal of Ethnographic Theory* 7 (1), 397–406.

Goleman, D. (2005) *Emotional Intelligence: Why It Can Matter More Than IQ*. New York: Bantam Books.

Gomez, P., Zimmermann, P., Guttormsen-Schar, S. and Danuser, B. (2005) Respiratory responses associated with affective processing of film stimuli. *Biological Psychology* 68 (12), 223–235.

Goodman, F.R., Disabato, D.J., Kashdan, T.B. and Kauffman, S.B. (2018) Measuring well-being: A comparison of subjective well-being and PERMA. *Journal of Positive Psychology* 13 (4), 321–332.

Goossens, C. (2000) Tourism information and pleasure motivation. *Annals of Tourism Research* 27, 301–321.

Gotham, K.F. (2007) *Authentic New Orleans: Tourism, Culture, and Race in the Big Easy*. New York: New York University Press.

Gothóni, R. (2000) The healing quality of pilgrimage to Mount Athos. *Archive for the Psychology of Religion* 23 (1), 132–143.

Gottlieb, A. (1982) Americans' vacations. *Annals of Tourism Research* 9 (2), 165–187.

Graham-McLay, C. (2019) A tourist family's bad behavior has New Zealand rethinking its welcome mat. *The New York Times*, 22 January. See https://www.nytimes.com/2019/01/22/world/asia/new-zealand-british-tourists.html (accessed September 2019).

Griffiths, J. (2019) Parks in Beijing want to blacklist 'uncivilized' visitors. *CNN Travel*, 8 April. See https://edition.cnn.com/travel/article/china-parks-tourism-blacklist/index.html/ (accessed September 2019).

Griffiths, M. (2016) Why would anyone want to have sex with an animal? The psychology of bestiality. *Independent*, 2 February. See https://www.independent.co.uk/life-style/love-sex/why-would-anyone-want-to-have-sex-with-an-animal-the-psychology-of-bestiality-10201158.html (accessed August 2019).

Grimwood, B.S. (2015) Advancing tourism's moral morphology: Relational metaphors for just and sustainable Arctic tourism. *Tourist Studies* 15 (1), 3–26.

Grimwood, B.S.R., Mair, H., Caton, K. and Muldoon, M. (eds) (2018) *Tourism and Wellness: Travel for the Good of All?* Lanham, MD: Rowman & Littlefield.

Gross, C., Piper, T.M., Bucciarelli, A., Tardiff, K., Vlahov, D. and Galea, S. (2007) Suicide tourism in Manhattan, New York City, 1990–2004. *Journal of Urban Health* 84 (6), 755–765.

Gunter, J. (2017) 'Yolocaust': How should you behave at a Holocaust memorial? *BBC News*, 20 January. See https://www.bbc.com/news/world-europe-38675835 (accessed November 2019).

Gursoy, D., Chi, C.G. and Dyer, P. (2009) Locals' attitudes toward mass and alternative tourism: The case of Sunshine Coast, Australia. *Journal of Travel Research* 49 (3), 381–394.

Guttentag, D.A. (2010) Virtual reality: Applications and implications for tourism. *Tourism Management* 31 (5), 637–651.

Haines, G. (2016) Is tourism in Thailand becoming unsustainable? *The Telegraph*, 17 May. See https://www.telegraph.co.uk/travel/news/is-tourism-in-thailand-becoming-unsustainable/ (accessed September 2019).

Hall, C.M. (2011) Health and medical tourism: A kill or cure for global public health? *Tourism Review* 66 (1–2), 4–15.

Hall, C.M. (ed.) (2013) *Medical Tourism: The Ethics, Regulation, and Marketing of Health Mobility*. London: Routledge.

Hall, C.M., Baird, T., James, M. and Ram, Y. (2016) Climate change and cultural heritage: Conservation and heritage tourism in the Anthropocene. *Journal of Heritage Tourism* 11 (1), 10–24.

Hallak, R., Brown, G. and Lindsay, N.J. (2012) The place identity–performance relationship among tourism entrepreneurs: A structural equation modelling analysis. *Tourism Management* 33, 143–154.

Hallinan, B. (2018a) 15 Beautiful sacred sites around the world. *Condé Nast Traveler*, 30 April. See https://www.cntraveler.com/gallery/beautiful-sacred-sites-around-the-world (accessed November 2019).

Hallinan, B. (2018b) New Zealand's Lake Wanaka Tree is being destroyed by tourists. *Condé Nast Traveler*, 23 January. See https://www.cntraveler.com/story/new-zealands-lake-wanaka-tree-is-being-destroyed-by-tourists (accessed October 2019).

Han, H., Back, K. and Barrett, B. (2009) Influencing factors on restaurant customers' revisit intention: The roles of emotions and switching barriers. *International Journal of Hospitality Management* 28 (4), 563–572.

Harrison, D. and Sharpley, R. (eds) (2017) *Mass Tourism in a Small World*. Wallingford: CABI.

Hart, T. (2000) Deep empathy. In T. Hart, P.L. Nelson and K. Puhakka (eds) *Transpersonal Knowing: Exploring the Horizon of Consciousness* (pp. 253–270). New York: State University of New York Press.

Hartmann, R. (2014) Dark tourism, thanatourism, and dissonance in heritage tourism management: New directions in contemporary tourism research. *Journal of Heritage Tourism* 9 (2), 166–182.

Hartwell, H., Fyall, A., Willis, C., Page, S., Ladkin, A. and Hemingway, A. (2018) Progress in tourism and destination wellbeing research. *Current Issues in Tourism* 21 (16), 1830–1892.

Hatfield, E. and Rapson, R. (1993a) *Love and Sex: Cross-cultural Perspectives*. Needham Heights, MA: Allyn & Bacon.

Hatfield, E. and Rapson, R. (1993b) Love and attachment processes. In M. Lewis and J. Haviland-Jones (eds) *Handbook of Emotions* (pp. 595–604). London: Guilford Press.

Hausmann, A., Slotow, R., Burns, J. and Di Minin, E. (2016) The ecosystem service of sense of place: Benefits for human well-being and biodiversity conservation. *Environmental Conservation* 43 (2), 117–127.

Hejmadi, A., Davidson, R. and Rozin, P. (2000) Exploring Hindu Indian emotion expressions. *Psychological Science* 11, 183–187.

Hellén, K. and Sääksjärvi, M. (2013) Development of a scale measuring childlike anthropomorphism in products. *Journal of Marketing Management* 29, 141–157.

Henderson, J.C. (2016) Halal food, certification and halal tourism: Insights from Malaysia and Singapore. *Tourism Management Perspectives* 19, 160–164.

Henderson, M.D., Huang, S. and Chang, C.A. (2012) When others cross psychological distance to help: Highlighting prosocial actions toward outgroups encourages philanthropy. *Journal of Experiential Social Psychology* 48 (1), 220–225.

Hernández, J.M., Suárez-Vega, R. and Santana-Jiménez, Y. (2016) The inter-relationship between rural and mass tourism: The case of Catalonia, Spain. *Tourism Management* 54, 43–57.

Hill, P.C., Pargament, K., Hood, R.W., McCullough, M.E. and Swyers, J.P. (2000) Conceptualising religion and spirituality: Points of commonality, points of departure. *Journal for the Theory of Social Behaviour* 30, 51–77.

Hindley, A. and Font, X. (2017) Ethics and influences in tourist perceptions of climate change. *Current Issues in Tourism* 20 (16), 1684–1700.

Hinsliff, G. (2018) Airbnb and the so-called sharing economy is hollowing out our cities. *The Guardian*, 31 August. See https://www.theguardian.com/commentisfree/2018/

aug/31/airbnb-sharing-economy-cities-barcelona-inequality-locals (accessed November 2019).
Höckert, E. (2015) *Ethics of Hospitality: Participatory Tourism Encounters in the Northern Highlands of Nicaragua*. Rovaniemi: Lapland University Press.
Höckert, E. (2018) *Negotiating Hospitality: Ethics of Tourism Development in the Nicaraguan Highlands*. Abingdon: Routledge.
Hodalska, M. (2017) Selfies at horror sites: Dark tourism, ghoulish souvenirs and digital narcissism. *Zeszyty Prasoznawcze* 230 (2), 405–423.
Holden, A. (2003) In need of new environmental ethics for tourism? *Annals of Tourism Research* 30 (1), 94–108.
Holden, A. (2009) The environment–tourism nexus: Influence of market ethics. *Annals of Tourism Research* 36 (3), 373–389.
Holden, A. (2018) Environmental ethics for tourism – the state of the art. *Tourism Review* 74 (3), 694–703.
Holjevac, I.A. (2008) Business ethics in tourism – as a dimension of TQM. *Total Quality Management and Business Excellence* 19 (10), 1029–1041.
Holy Hesychasterion (2018) *Saint Paisios the Athonite*. Vasilika: Holy Hesychasterion 'Evangelist John the Theologian'.
Homer (2004) *Homer's Iliad*. Thessaloniki: Zitros [in Greek].
Hosany, S. (2011) Appraisal determinants of tourist emotional responses. *Journal of Travel Research* 51 (3), 303–314.
Hosany, S. and Gilbert, D. (2009) Measuring tourists' emotional experiences toward hedonic holiday destinations. *Journal of Travel Research* 49 (4), 513–526.
Hosany, S., Prayag, G., Van der Veen, R., Huang, S. and Deesilatham, S. (2016) Mediating effects of place attachment and satisfaction on the relationship between tourists' emotions and intention to recommend. *Journal of Travel Research* 56 (8), 1079–1093.
Hu, H.H.S., Hu, H.Y. and King, B. (2017) Impacts of misbehaving air passengers on frontline employees: Role stress and emotional labor. *International Journal of Contemporary Hospitality Management* 29 (7), 1793–1813.
Hu, H.W. and Yoshikawa, T. (2017) CEO and board influence on corporate philanthropy in China. *Academy of Management Proceedings* 1, 12453.
Huang, S.-C.L. (2013) Visitor responses to the changing character of the visual landscape as an agrarian area becomes a tourist destination: Yilan County, Taiwan. *Journal of Sustainable Tourism* 21 (1), 154–171.
Hudson, S. and Miller, G. (2005) Ethical orientation and awareness of tourism students. *Journal of Business Ethics* 62 (4), 383–396.
Hudson, S., Roth, M.S., Madden, T.J. and Hudson, R. (2015) The effects of social media on emotions, brand relationship quality, and word of mouth: An empirical study of music festival attendees. *Tourism Management* 47, 68–76.
Hughes, P. (2001) Animals, values and tourism – structural shifts in UK dolphin tourism provision. *Tourism Management* 22 (4), 321–329.
Hughes, R. (2016) Do tourists really go to Afghanistan? *BBC News*, 4 August. See https://www.bbc.com/news/world-asia-36974513 (accessed August 2019).
Hull, E.M. (2011) Sex, drugs and gluttony: How the brain controls motivated behaviors. *Physiology & Behavior* 104, 173–177.
Hultman, M., Skarmeas, D., Oghazi, P. and Beheshti, H.M. (2015) Achieving tourist loyalty through destination personality, satisfaction, and identification. *Journal of Business Research* 68 (11), 2227–2231.
Hunt, E. (2018) Residents in tourism hotspots have had enough. So what's the answer? *The Guardian*, 17 July. See https://www.theguardian.com/cities/2018/jul/17/residents-in-tourism-hotspots-have-had-enough-so-whats-the-answer (accessed September 2019).
Hutchinson, D.S. (2015) *The Virtues of Aristotle*. New York: Routledge.

Hutt, R. (2019) Sweden is a top performer on well-being. Here's why. *World Economic Forum*, 31 May. See https://www.weforum.org/agenda/2019/05/sweden-is-a-top-performer-on-well-being-here-s-why/ (accessed October 2019).
Hyde, K.F. and Harman, S. (2011) Motives for a secular pilgrimage to the Gallipoli battlefields. *Tourism Management* 32 (6), 1343–1351.
Innerarity, D. (2017) *Ethics of Hospitality*. London: Taylor & Francis.
Io, M.U. (2016) Exploring the impact of hedonic activities on casino-hotel visitors' positive emotions and satisfaction. *Journal of Hospitality and Tourism Management* 26, 27–35.
Ioannides, D. (2008) Hypothesizing the shifting mosaic of attitudes through time: A dynamic framework for sustainable tourism development on a 'Mediterranean Isle'. In S. McCool and R.N. Moisey (eds) *Tourism, Recreation and Sustainability: Linking Culture and the Environment* (pp. 51–75). Wallingford: CABI.
Ioannides, D. and Holcomb, B. (2003) Misguided policy initiatives in small-island destinations: Why up-market tourism policies fail? *Tourism Geographies* 5 (1), 39–48.
Ioannides, M.W.C. and Ioannides, D. (2006) Global Jewish tourism: Pilgrimages and remembrance. In D. Timothy and D.H. Olsen (eds) *Tourism, Religion and Spiritual Journeys* (pp. 156–171). London: Routledge.
Irakleous, C. (2015) *Paraklisi*. Lemesos: Iera Mitropoli Lemesou.
Irvine, W.B. (2010) On gluttony: Religious and philosophical responses to the obesity epidemic. In L. Dube, A. Bechara, A. Dagher, A. Drewnowski, J. LeBel, P. James, D. Richard and R.Y. Yada (eds) *Obesity Prevention: The Role of Brain and Society on Individual Behavior* (pp. 653–660). Amsterdam: Elsevier.
Isaac, E. (2004) *Life of Elder Paisios the Hagiorite*. Thessaloniki: Agion Oros.
Isaac, H. (2016) *Saint Paisios of Mount Athos* (2nd edn). Chalkidiki: Holy Monastery of Saint Arsenios the Cappadocian.
Isacsson, A., Alakoski, L. and Bäck, A. (2009) Using multiple senses in tourism marketing: The Helsinki expert, Eckerö Line and Linnanmäki amusement park cases. *Tourismos: An International Multidisciplinary Journal of Tourism* 4 (3), 167–184.
Isaak, S. (2011) *Maxims of Abba Isaak the Syrian*. Thessaloniki: Enomeni Romiosini.
Izard, C.E. (1971) *The Face of Emotion*. New York: Appleton-Century-Crofts.
Izard, C.E. (1978) *Human Emotions*. New York: Springer Science and Business Media.
Jabreel, M., Moreno, A. and Huertas, A. (2017) Semantic comparison of the emotional values communicated by destinations and tourists on social media. *Journal of Destination Marketing and Management* 6 (3), 170–183.
Jamal, T.B. (2004) Virtue ethics and sustainable tourism pedagogy: Phronesis, principles and practice. *Journal of Sustainable Tourism* 12 (6), 530–545.
James, W. (1884) What is an emotion? *Mind* 9, 188–205.
James, W. (1894) The physical basis of emotion. *Psychological Review* 1, 516–529.
Jamrozy, U. and Uysal, M. (1994) Travel motivation variations of overseas German visitors. *Global Tourist Behaviour* 6 (3–4), 135–160.
Jang, S.S. and Wu, C.M. (2006) Seniors' travel motivation and the influential factors: An examination of Taiwanese seniors. *Tourism Management* 27 (2), 306–316.
Japan National Tourism Organization (2019) *Kansai: Kyoto – The Old Imperial Capital and Cultural Heart of Japan*. See https://www.japan.travel/en/destinations/kansai/kyoto/ (accessed August 2019).
Jarratt, D. and Gammon, S. (2016) We had the most wonderful times: Seaside nostalgia at a British resort. *Tourism Recreation Research* 41, 123–133.
Jaszay, C. (2002) Teaching ethics in hospitality programs. *Journal of Hospitality and Tourism Education* 14 (3), 57–63.
Jeong, S. and Santos, C.A. (2004) Cultural politics and contested place identity. *Annals of Tourism Research* 31 (3), 640–656.

Jepson, D. and Sharpley, R. (2015) More than sense of place? Exploring the emotional dimension of rural tourism experiences. *Journal of Sustainable Tourism* 23 (8–9), 1157–1178.

Jiménez Beltrán, J., López-Guzmán, T. and Santa-Cruz, F.G. (2016) Gastronomy and tourism: Profile and motivation of international tourism in the city of Córdoba, Spain. *Journal of Culinary Science and Technology* 14 (4), 347–362.

John Climacus (2019) *The Ladder of Divine Ascent* (A. Lazarus Moore, trans.). New York: Harper.

Johnston, T. (2015) The geographies of thanatourism. *Geography* 100, 20.

Jordan, E.J., Spencer, D.M. and Prayag, G. (2019) Tourism impacts, emotions and stress. *Annals of Tourism Research* 75, 213–226.

Jorgensen, B.S. and Stedman, R.C. (2006) A comparative analysis of predictors of sense of place dimensions: Attachment to, dependence on, and identification with lakeshore properties. *Journal of Environmental Management* 79, 316–327.

Jovicic, D.Z. (2014) Key issues in the implementation of sustainable tourism. *Current Issues in Tourism* 17 (4), 297–302.

Jung, T., tom Dieck, M.C., Lee, H. and Chung, N. (2016) Effects of virtual reality and augmented reality on visitor experiences in museum. In A. Inversini and R. Schegg (eds) *Information and Communication Technologies in Tourism 2016* (pp. 621–635). Cham: Springer.

Jungsik, K. and Hatfield, E. (2004) Love types and subjective well-being: A cross-cultural study. *Social Behavior and Personality: An International Journal* 32 (2), 173–182.

Kaaristo, M. (2014) Value of silence: Mediating aural environments in Estonian rural tourism. *Journal of Tourism and Cultural Change* 12 (3), 267–279.

Kalat, J.W. (2011) *Introduction to Psychology* (9th edn). Belmont, CA: Wadsworth Cengage Learning.

Kaplan, H.A. (1987) The psychopathology of nostalgia. *Psychoanalytic Review* 74 (4), 465–486.

Katanich, D. (2017) Five destinations where luxury and philanthropy go hand in hand. *Euronews*, 22 August. See https://www.euronews.com/living/2017/08/22/five-destinations-where-luxury-and-philanthropy-go-hand-in-hand# (accessed August 2019).

Kato, K. and Progano, R.N. (2017) Spiritual (walking) tourism as a foundation for sustainable destination development: Kumano-kodo pilgrimage, Wakayama, Japan. *Tourism Management Perspectives* 24, 243–251.

Katsaitis, A. and Papaefthimiou, E. (2019) Philoxenia as a component of the tourism experience in culture and total quality management in hotel sector. In V. Katsoni and M. Segarra-Oña (eds) *Smart Tourism as a Driver for Culture and Sustainability* (207–221) Cham: Springer.

Kaufman, W.R. (2007) Gluttony and sex in female ixodid ticks: How do they compare to other blood-sucking arthropods? *Journal of Insect Physiology* 53 (3), 264–273.

Keating, B. (2009) Managing ethics in the tourism supply chain: The case of Chinese travel to Australia. *International Journal of Tourism Research* 11 (4), 403–408.

Kelloway, E., Inness, M., Barling, J., Francis, L. and Turner, N. (2010) Loving one's job: Construct development and implications for individual well-being. In P. Perrewe and C. Ganster (eds) *New Developments in Theoretical and Conceptual Approaches to Job Stress* (pp. 109–136). Bingley: Emerald Group.

Kelly, C. (2012) Wellness tourism: Retreat visitor motivations and experiences. *Tourism Recreation Research* 37 (3), 205–213.

Kelly, R., Losekoot, E. and Wright-StClair, V.A. (2016) Hospitality in hospitals: The importance of caring about the patient. *Hospitality & Society* 6 (2), 113–129.

Kemp, S., Burt, C.D.B. and Furneaux, L. (2008) A test of the peak-end rule with extended autobiographical events. *Memory and Cognition* 36 (1), 132–138.

Kennedy, D. (1998) Shakespeare and cultural tourism. *Theatre Journal* 50 (2), 175–188.

Kerstetter, D. and Bricker, K. (2009) Exploring Fijian's sense of place after exposure to tourism development. *Journal of Sustainable Tourism* 6, 691–708.

Kessous, A. and Roux, E. (2008) A semiotic analysis of nostalgia as a connection to the past. *Qualitative Market Research: An International Journal* 11 (2), 192–212.

Khan, M. (2019) Relationship of ethical behavior with material well-being and happiness: An Islamic perspective. doi:10.13140/RG.2.2.13513.70240

Khazaee-Pool, M., Sadeghi, R., Majlessi, F. and Rahimi Foroushani, A. (2015) Effects of physical exercise programme on happiness among older people. *Journal of Psychiatric and Mental Health Nursing* 22 (1), 47–57.

Kilduff, M., Chiaburu, D.S. and Menges, J.I. (2010) Strategic use of emotional intelligence in organizational settings: Exploring the dark side. *Research in Organizational Behavior* 30, 129–152.

Kilipiris, F.E. and Dermetzopoulos, A. (2016) Streets of Orthodoxy: Developing religious tourism in the Mount Paiko area, Central Macedonia, Greece. *International Journal of Religious Tourism and Pilgrimage* 4 (7), 31–37.

Kim, J.J. and Fesenmaier, D.R. (2014) Measuring emotions in real time: Implications for tourism experience design. *Journal of Travel Research* 54 (4), 419–429.

Kim, J.J. and Fesenmaier, D.R. (2015) Designing tourism places: Understanding the tourism experience through our senses. *Advancing Tourism Research Globally* 19.

Kim, J. and Fesenmaier, D.R. (2017) Sharing tourism experiences: The posttrip experience. *Journal of Travel Research* 56 (1), 28–40.

Kim, J. and James, J.D. (2019) Sport and happiness: Understanding the relations among sport consumption activities, long-and short-term subjective well-being, and psychological need fulfillment. *Journal of Sport Management* 33 (2), 119–132.

Kim, S., Park, E. and Lamb, D. (2019) Extraordinary or ordinary? Food tourism motivations of Japanese domestic noodle tourists. *Tourism Management Perspectives* 29, 176–186.

Kinnamon, M. (2015) Koinonia and philoxenia: Toward an expanded ecumenical ecclesiology. *Ecumenical Trends* 44 (10), 1–5.

Kirchgaessner, S. (2015) Consumer affairs: Top tips on tipping around the world. *The Guardian*, 25 July. See https://www.theguardian.com/money/2015/jul/25/tipping-around-the-world-holiday-restaurants-taxi-drivers-what-pay (accessed August 2019).

Kivela, J. and Crotts, J.C. (2006) Tourism and gastronomy: Gastronomy's influence on how tourists experience a destination. *Journal of Hospitality and Tourism Research* 30 (3), 354–377.

Kivotos (2019) *Saint Nektarios: The Road to Happiness.* See (in Greek) http://ikivotos.gr/post/8585/o-dromos-ths-eytyxias# (accessed October 2019).

Knani, M. (2014) Ethics in the hospitality industry: Review and research agenda. *International Journal of Business and Management* 9 (3), 1–8.

Knapton, S. (2017) Your brain on nostalgia: First study shows neurons light up in meaningful places. *The Telegraph*, 12 October. See https://www.telegraph.co.uk/science/2017/10/12/brain-nostalgia-first-study-shows-neurons-light-meaningful-places/ (accessed September 2019).

Knaus, C. (2017) The race to rescue Cambodian children from orphanages exploiting them for profit. *The Guardian*, 18 August. See https://www.theguardian.com/world/2017/aug/19/the-race-to-rescue-cambodian-children-from-orphanages-exploiting-them-for-profit (accessed August 2019).

Knudsen, B.T. (2011) Thanatourism: Witnessing difficult pasts. *Tourist Studies* 11 (1), 55–72.

Knudsen, B. and Waade, A. (2010) *Re-Investing Authenticity: Tourism, Place and Emotions.* Bristol: Channel View Publications.

Koc, E. (2013) Inversionary and liminoidal consumption: Gluttony on holidays and obesity. *Journal of Travel & Tourism Marketing* 30 (8), 825–836.
Koc, E. (2016) Food consumption in all-inclusive holidays: Illusion of control as an antecedent of inversionary consumption. *Journal of Gastronomy and Tourism* 2 (2), 107–116.
Kollmann, J. (1885) Das ueberwintern von Europaischen frosch- und tritonlarven und die umwandlung des Mexikanischen axolotl. *Verhandlungen d. Naturforschende Gesellschaft (Basel)* 7, 387–39.
Komppula, R., Konu, H. and Vikman, N. (2017) Listening to the sounds of silence: Forest based wellbeing tourism in Finland. In J. Chen and N. Prebensen (eds) *Nature Tourism: A Global Perspective* (pp. 120–130). Abingdon: Routledge.
Kontogeorgopoulos, N. (2017) Finding oneself while discovering others: An existential perspective on volunteer tourism in Thailand. *Annals of Tourism Research* 65, 1–12.
Konu, H. (2015) Case study: Developing a forest-based wellbeing tourism product together with customers – an ethnographic approach. *Tourism Management* 49, 1–16.
Kosmin, B.A. and Ritterband, P. (eds) (1991) *Contemporary Jewish Philanthropy in America*. Lanham, MD: Rowman & Littlefield.
Kostel Sv. Mikuláše (2019) *Sightseeing*. See http://www.stnicholas.cz/en/sightseeing/sightseeing---entrance-fee/ (accessed November 2019).
Koster, W. (2019) Bali tells tourists: Violate our way of life and we'll send you home. *Traveller*, 19 August. See http://www.traveller.com.au/bali-violate-our-way-of-life-and-well-send-you-home-bali-tells-tourists-h1h8ck (accessed August 2019).
Kotsonis, I. (1997) *An Athonite Gerontikon: Sayings of the Holy Fathers of Mount Athos* (M.D. Mayson and T. Zion, trans.). Thessaloniki: Publications of the Holy Monastery of St. Gregory Palamas.
Kozak, M. and Rimmington, M. (2000) Tourist satisfaction with Mallorca, Spain, as an off-season holiday destination. *Journal of Travel Research* 38 (3), 260–269.
Kozub, K.R., O'Neill, M.A. and Palmer, A.A. (2014) Emotional antecedents and outcomes of service recovery. *Journal of Services Marketing* 28 (3), 233–243.
Kraut, R. (1991) *Aristotle on the Human Good*. Princeton, NJ: Princeton University Press.
Kraye, J. (2007) The revival of Hellenistic philosophies. In J. Hankins (ed.) *The Cambridge Companion to Renaissance Philosophy* (pp. 97–112). Cambridge: Cambridge University Press.
Kringelbach, M.L. and Berridge, K.C. (2017) The affective core of emotion: Linking pleasure, subjective well-being, and optimal metastability in the brain. *Emotion Review* 9 (3), 191–199.
Kristjánsson, K. (2016) *Aristotle, Emotions, and Education*. Abingdon: Routledge.
Krueger, A. (2019) It's all in the (fine) details for upscale hotels seeking to elevate guest experiences. *Washington Post*, 2 August. See https://www.washingtonpost.com/lifestyle/travel/its-all-in-the-fine-details-for-upscale-hotels-seeking-to-elevate-guest-experiences/2019/08/01/6a6b02c4-9f3d-11e9-b27f-ed2942f73d70_story.html (accessed August 2019).
Kuo, C.M., Chen, L.C. and Tseng, C.Y. (2017) Investigating an innovative service with hospitality robots. *International Journal of Contemporary Hospitality Management* 29 (5), 1305–1321.
Lai, I.K.W. and Hitchcock, M. (2017) Local reactions to mass tourism and community tourism development in Macau. *Journal of Sustainable Tourism* 25 (4), 451–470.
Laing, J.H. and Crouch, G.I. (2011) Frontier tourism: Retracing mythic journeys. *Annals of Tourism Research* 38 (4), 1516–1534.
Lange, C.G. and James, W. (1967) *The Emotions*. New York: Hafner.
Lange, G. (2015) Tourism in Zanzibar: Incentives for sustainable management of the coastal environment. *Ecosystem Services* 11, 5–11.
Lantos, G.P. (2002) The ethicality of altruistic corporate social responsibility. *Journal of Consumer Marketing* 19 (3), 205–232.

Lashley, C. (2008) Marketing hospitality and tourism experiences. In H. Oh (ed.) *Handbook of Hospitality Marketing Management* (pp. 4–31). Oxford: Butterworth-Heinemann.
Lashley, C. (2014) Insights into the morality of hospitality. *Hospitality & Society* 4 (1), 3–7.
Lashley, C. (2015a) Hospitality experience: An introduction to hospitality management. *Journal of Tourism Futures* 1 (2), 160–161.
Lashley, C. (2015b) Hospitality and hospitableness. *Research in Hospitality Management* 5 (1), 1–7.
Lashley, C. (ed.) (2016) *The Routledge Handbook of Hospitality Studies*. Abingdon: Routledge.
Lashley, C. (2017) Introduction: Research on hospitality. The story so far/ways of knowing. In C. Lashley (ed.) *The Routledge Handbook of Hospitality Studies* (pp. 1–10). Abingdon: Routledge.
Lashley, C. and Lynch, P. (2013) Control and hospitality. *Hospitality & Society* 3 (1), 3–6.
Lazari-Radek, K. and Singer, P. (2017) *Utilitarianism: A Very Short Introduction*. Oxford: Oxford University Press.
Lazarus, R. (1991) *Emotion and Adaptation*. New York: Oxford University Press.
Lazarus, R.S. (1999) Hope: An emotion and a vital coping resource against despair. *Social Research* 66 (2), 653–678.
Lazarus (Moore), Archimandrite (2009) *An Extraordinary Peace: St Seraphim, Flame of Sarov*. Port Townsend: Anaphora Press.
Lecompte, A.F., Trelohan, M., Gentric, M. and Aquilina, M. (2017) Putting sense of place at the centre of place brand development. *Journal of Marketing Management* 33 (5–6), 400–420.
Lee, J. (1977) A typology of styles of loving. *Personality Social Psychological Bulletin* 3 (2), 173–182.
Lee, J.J. and Kyle, G.T. (2011) Recollection consistency of festival consumption emotions. *Journal of Travel Research* 51 (2), 178–190.
Lee, T.J. and Kim, J.S. (2016) Relationships between emotion regulation seeking, programme satisfaction, attention restoration and life satisfaction: Healing programme participants. In M.K. Smith and L. Puczko (eds) *Routledge Handbook of Health Tourism* (pp. 375–385). London: Routledge.
Lee, Y., Choi, J., Moon, B. and Babin, B. (2014) Codes of ethics, corporate philanthropy, and employee responses. *International Journal of Hospitality Management* 39, 97–106.
Legorano, G. (2020) Venice's high tides receded. Now tourism is running dry. *The Wall Street Journal*, 1 March. See https://www.wsj.com/articles/venices-high-tides-receded-now-tourism-is-running-dry-11583066058/ (accessed March 2020).
Leisinger, K.M. (2007) Corporate philanthropy: The 'top of the pyramid'. *Business and Society Review* 112 (3), 315–342.
Lennon, J. (2010) Dark tourism and sites of crime. In D. Botterill and T. Jones (eds) *Tourism and Crime: Key Themes*. Oxford: Goodfellow.
Lennon, M. (2019) 10 secrets Disney parks staff don't want you to know. *The Travel*, 23 August. See https://www.thetravel.com/disney-park-staff-secrets/ (accessed August 2019).
Lepp, A. and Gibson, H. (2008) Sensation seeking and tourism: Tourist role, perception of risk and destination choice. *Tourism Management* 29 (4), 740–750.
Levi, E., Dolev, T., Colins-Kreiner, N. and Zilcha-Mano, S. (2019) Tourism and depressive symptoms. *Annals of Tourism Research* 74, 191–194.
Levine, L. and Pizzaro, D. (2004) Emotion and memory research: A grumpy overview. *Social Cognition* 5 (12), 530–554.

Levine, M. and Thompson, K. (2004) Identity, place, and bystander intervention: Social categories and helping after natural disasters. *Journal of Social Psychology* 144, 229–245.

Lewis, J. (2019) The Notre Dame fire has a lesson for philanthropy. *National Committee for Responsive Philanthropy*, 18 April. See https://www.ncrp.org/2019/04/the-notre-dame-fire-has-a-lesson-for-philanthropy.html (accessed August 2019).

Liamis, E. (2019) *Tourists and Visitors*. See https://pemptousia.com/ (accessed August 2019).

Light, D. (2009) Performing Transylvania: Tourism, fantasy and play in a liminal space. *Tourist Studies* 9 (3), 240–258.

Light, D. (2017) Progress in dark tourism and thanatourism research: An uneasy relationship with heritage tourism. *Tourism Management* 61, 275–301.

Lin, Y., Kerstetter, D., Nawijn, J. and Mitas, O. (2014) Changes in emotions and their interactions with personality in a vacation context. *Tourism Management* 40, 416–424.

Lindner, A.M. (2016) The United States of excess: Gluttony and the dark side of American exceptionalism. *Contemporary Sociology* 45 (4), 490–492.

Linenthal, E.T. (2001) *Preserving Memory: The Struggle to Create America's Holocaust Museum*. New York: Columbia University Press.

Liu, B., Floud, S., Pirie, K., Green, J., Peto, R., Beral, V. and Million Women Study Collaborators (2016) Does happiness itself directly affect mortality? The prospective UK Million Women Study. *The Lancet* 387 (10021), 874–881.

Liu, J. (2016) Hotels recycle soaps for charity. *BBC News*, 8 July. See https://www.bbc.com/news/av/business-36742211/hotels-recycle-soaps-for-charity (accessed September 2019).

Liu, S. and Cheung, L.T. (2016) Sense of place and tourism business development. *Tourism Geographies* 18 (2), 174–193.

Lo, A., Wu, C. and Tsai, H. (2015) The impact of service quality on positive consumption emotions in resort and hotel spa experiences. *Journal of Hospitality Marketing & Management* 24 (2), 155–179.

Lois-González, R.C. and Santos, X.M. (2015) Tourists and pilgrims on their way to Santiago. Motives, Caminos and final destinations. *Journal of Tourism and Cultural Change* 13 (2), 149–164.

Lonely Planet (2012) Top 10 hedonistic city breaks. *Lonely Planet*, 20 June. See https://www.lonelyplanet.com/travel-tips-and-articles/top-10-hedonistic-city-breaks/40625c8c-8a11-5710-a052-1479d2777dba (accessed August 2019).

López Díaz, A. (2017) Why Barcelona locals really hate tourists. *Independent*, 9 August. See https://www.independent.co.uk/travel/news-and-advice/barcelona-locals-hate-tourists-why-reasons-spain-protests-arran-airbnb-locals-attacks-graffiti-a7883021.html (accessed November 2019).

López-Guzmán, T. and Sánchez-Cañizares, S. (2012) Culinary tourism in Córdoba (Spain). *British Food Journal* 114 (2), 168–179.

López-Guzmán, T., Orgaz-Agüera, F., Martín, J.A.M. and Ribeiro, A. (2016) The all-inclusive tourism system in Cape Verde islands: The tourists' perspective. *Journal of Hospitality and Tourism Management* 29, 9–16.

Lovelock, B. and Lovelock, K. (2013) *The Ethics of Tourism: Critical and Applied Perspectives*. London: Routledge.

Lucas, J. (2019) 'Killing them with kindness': Fear for kangaroos at WA's Lucky Bay as tourists seek selfies. *ABC News*, 11 April. See https://www.abc.net.au/news/2019-04-12/wildlife-authorities-hopping-mad-over-tourists-feeding-kangaroos/10994880 (accessed August 2019).

Lucas, R.E. (2007) Adaptation and the set-point model of subjective well-being: Does happiness change after major life events? *Current Directions in Psychological Science* 16 (2), 75–79.

Lugosi, P., Robinson, R.N.S., Golubovskaya, M. and Foley, L. (2016) The hospitality consumption experiences of parents and carers with children: A qualitative study of foodservice settings. *International Journal of Hospitality Management* 54, 84–94.
Luong, D. (2015) Vietnam's rush to develop risks damaging its natural attractions. *The Guardian*, 1 August. See https://www.theguardian.com/world/2015/aug/01/vietnam-tourism-rush-development-conservation (accessed October 2019).
Lynch, A.P., Molz, J.G., Mcintosh, A., Lugosi, P. and Lashley, C. (2011) Theorizing hospitality. *Hospitality & Society* 1 (1), 3–24.
Lyubomirksy, S., Sheldon, K.M. and Schkade, D. (2005) Pursuing happiness: The architecture of sustainable change. *Review of General Psychology* 9 (2), 111–131.
Machuca, D.E. (2006) The Pyrrhonist's ἀταραξία and φιλανθρωπία. *Ancient Philosophy* 26 (1), 111–139.
Mackenzie, S. and Kerr, J.H. (2013) Stress and emotions at work: An adventure tourism guide's experiences. *Tourism Management* 36, 3–14.
MacNeil, T. and Wozniak, D. (2018) The economic, social, and environmental impacts of cruise tourism. *Tourism Management* 66, 384–404.
Macpherson, F. (ed.) (2011) *The Senses: Classic and Contemporary Philosophical Perspectives*. Oxford: Oxford University Press.
Macris, V. (2012) Towards a pedagogy of philoxenia (hospitality): Negotiating policy priorities for immigrant students in Greek public schools. *Journal for Critical Education Policy Studies (JCEPS)* 10 (1).
Madera, J.M., Neal, J.A. and Dawson, M. (2011) A strategy for diversity training: Focusing on empathy in the workplace. *Journal of Hospitality and Tourism Research* 35 (4), 469–487.
Mak, A.H.N., Lumbers, M., Eves, A. and Chang, R.C.Y. (2013) An application of the repertory grid method and generalised procrustes analysis to investigate the motivational factors of tourist food consumption. *International Journal of Hospitality Management* 35, 327–338.
Mak, A.H.N., Lumbers, M., Eves, A. and Chang, R.C.Y. (2016) The effects of food-related personality traits on tourist food consumption motivations. *Asia Pacific Journal of Tourism Research* 22 (1), 1–20.
Mala, E. (2019) Cow cuddling: The latest therapeutic trend taking off in the US. *Independent*, 20 July. See https://www.independent.co.uk/arts-entertainment/cow-cuddling-america-netherlands-therapy-a9007096.html (accessed August 2019).
Malone, S., McCabe, S. and Smith, A.P. (2014) The role of hedonism in ethical tourism. *Annals of Tourism Research* 44, 241–254.
Malone, S., McKechnie, S. and Tynan, C. (2017) Tourists' emotions as a resource for customer value creation, cocreation, and destruction: A customer-grounded understanding. *Journal of Travel Research* 57 (7), 845–855.
Mantzarides, G.I. (2005) *Odoiporiko of Theological Anthropology*. Karyes: Holy Great Monastery of Vatopedi.
March, R. (1994) Tourism marketing myopia. *Tourism Management* 15 (6), 411–415.
Marchoo, W., Butcher, K. and Watkins, M. (2014) Tour booking: Do travelers respond to tourism accreditation and codes of ethics initiatives? *Journal of Travel & Tourism Marketing* 31 (1), 16–36.
Markides, K.C. (2017) The healing spirituality of Eastern Orthodoxy: A personal journey of discovery. *Religions* 8 (6), 109.
Marnburg, E. (2006) 'I hope it won't happen to me!': Hospitality and tourism students' fear of difficult moral situations as managers. *Tourism Management* 27 (4), 561–575.
Marschall, S. (2012) Tourism and memory. *Annals of Tourism Research* 39 (4), 2216–2219.

Martins, J., Gonçalves, R., Branco, F., Barbosa, L., Melo, M. and Bessa, M. (2017) A multisensory virtual experience model for thematic tourism: A Port wine tourism application proposal. *Journal of Destination Marketing and Management* 6 (2), 103–109.

Matheson, C.M., Rimmer, R. and Tinsley, R. (2014) Spiritual attitudes and visitor motivations at the Beltane Fire Festival, Edinburgh. *Tourism Management* 44, 16–33.

Mathews, A. and Mackintosh, B. (2004) Take a closer look: Emotion modifies the boundary extension effect. *Emotion* 4 (12), 36–45.

Maxouris, C. (2019) This bed and breakfast's 'cow cuddling' option may be exactly what we all need in life. *CNN*, 24 July. See https://edition.cnn.com/2019/07/24/us/cow-cuddling-bed-and-breakfast-trnd/index.html (accessed August 2019).

McCabe, S. and Johnson, S. (2013) The happiness factor in tourism: Subjective well-being and social tourism. *Annals of Tourism Research* 42, 42–65.

McCann, J.T., Graves, D. and Cox, L. (2014) Servant leadership, employee satisfaction, and organisational performance in rural community hospitals. *International Journal of Business and Management* 9 (10), 28–41.

McCarthy, J. (2016) The psychology of spa: The science of 'holistic' wellbeing. In M.K. Smith and L. Puczko (eds) *The Routledge Handbook of Health Tourism* (pp. 127–137). London: Routledge.

McGuire, C. (2016) The rudest place on earth? Shanghai Disneyland visitor films fellow tourists urinating in the streets, vandalising the theme park and pushing each other around. *Daily Mail*, 19 July. See https://www.dailymail.co.uk/travel/travel_news/article-3695827/Urinating-streets-vandalised-light-sockets-pushy-crowds-Video-shows-rude-tourists-Shanghai-s-Disneyland.html (accessed September 2019).

McTighe, K. (1984) Socrates on Desire for the Good and the Involuntariness of Wrongdoing: 'Gorgias' 466a-468e. *Phronesis* 193–236.

McVeigh, T. (2009) Tourist hordes told to stay away from world heritage sites by the locals. *The Guardian*, 6 September. See https://www.theguardian.com/environment/2009/sep/06/mass-tourism-environmental-damage (accessed November 2019).

Mehl, M.R., Vazire, S., Holleran, S.E. and Clark, C.S. (2010) Eavesdropping on happiness: Well-being is related to having less small talk and more substantive conversations. *Psychological Science* 21, 539–541.

Mesa Potamos (2019) *The Romanov Royal Martyrs: What Silence Could Not Conceal.* Moniatis: Mesa Potamos Monastery. See https://www.romanovs.eu.

Meyer, J.L. (2003) *The Spirit of Yellowstone.* Lanham, MD: Roberts Rinehart.

Miesler, L., Leder, H. and Hermann, A. (2011) Isn't it cute: An evolutionary perspective of baby-schema effects on visual product designs. *International Journal of Design* 5, 17–30.

Milano, C., Cheer, J.M. and Novelli, M. (2019) *Overtourism.* Wallingford: CABI.

Miles, S. (2014) Battlefield sites as dark tourism attractions: An analysis of experience. *Journal of Heritage Tourism* 9 (2), 134–147.

Mill, J.S. (1998 [1863]) *Utilitarianism* (R. Crisp, ed.). Oxford: Oxford University Press.

Miller, D.S. and Gonzalez, C. (2013) When death is the destination: The business of death tourism – despite legal and social implications. *International Journal of Culture, Tourism and Hospitality Research* 7 (3), 293–306.

Miller, W.I. (1997) Gluttony. *Representations* 60, 90–112.

Millward, D. (2018) The rise of the robot butler – fad or the future of hotel room service? *The Telegraph*, 19 October. See https://www.telegraph.co.uk/travel/hotels/articles/hotel-robot-room-service/ (accessed August 2019).

Mitas, O., Yarnal, C. and Chick, G. (2012) Jokes build community: Mature tourists' positive emotions. *Annals of Tourism Research* 39 (4), 1884–1905.

Mitchell, T.R., Thompson, L., Peterson, E. and Cronk, R. (1997) Temporal adjustments in the evaluation of events: The 'rosy view'. *Journal of Experimental Social Psychology* 33 (4), 421–448.

Moal-Ulvoas, G. (2017) Positive emotions and spirituality in older travelers. *Annals of Tourism Research* 66, 151–158.

Mody, M., Suess, C. and Lehto, X. (2019) Going back to its roots: Can hospitableness provide hotels competitive advantage over the sharing economy? *International Journal of Hospitality Management* 76 (A), 286–298.

Molloy, M. (2017) 'Traveling Butts' instagrammers arrested in Thailand over nude photo at historic temple. *The Telegraph*, 29 November. See https://www.telegraph.co.uk/news/2017/11/29/us-tourists-arrested-thailand-nude-photo-historic-temple/ (accessed November 2019).

Moloney, A. (2017) Child sex tourists do 'dirty business' with impunity in Dominican Republic. *Reuters*, 16 June. See https://www.reuters.com/article/us-dominican-sex-crimes/child-sex-tourists-do-dirty-business-with-impunity-in-dominican-republic-idUSKBN19727B (accessed August 2019).

Monks, K. (2016) Andrew Drury vacations in war zones – Iraq, Afghanistan, Somalia. *CNN Travel*, 12 January. See https://edition.cnn.com/travel/article/disaster-war-zone-dark-tourism-andrew-drury/index.html (accessed November 2019).

Moore, S. (2017) I don't mean to ruin your holiday, but Europe hates tourists – and with good reason. *The Guardian*, 16 August. See https://www.theguardian.com/commentisfree/2017/aug/16/dont-mean-ruin-your-holiday-but-europe-hates-tourists-with-good-reason-suzanne-moore/ (accessed November 2019).

Morley, I.B. (2009) The contemporary Chinese metropolis: Modernity, globalisation, and conceptual meanings. *Design Principles and Practices: An International Journal* 3 (1), 309–322.

Morris, H. (2018a) America's tipping culture is out of control – why should visitors be forced to reward bad service? *The Telegraph*, 15 June. See https://www.telegraph.co.uk/travel/comment/tipping-in-american-out-of-hand/ (accessed August 2019).

Morris, H. (2018b) Boozy lads no longer welcome in Ayia Napa, mayor warns. *The Telegraph*, 9 March. See https://www.telegraph.co.uk/travel/destinations/europe/cyprus/articles/ayia-napa-clamp-down-lads-holidays/ (accessed September 2019).

Moscardo, G. (2010) *Tourism Research Ethics: Current Considerations and Future Options*. Oxford: Goodfellow.

Moscardo, G.M. and Pearce, P.L. (1986) Historic theme parks: An Australian experience in authenticity. *Annals of Tourism Research* 13 (3), 467–479.

Mostafanezhad, M. and Hannam, K. (2016) *Moral Encounters in Tourism*. Abingdon: Routledge.

Moufakkir, O. and Selmi, N. (2018) Examining the spirituality of spiritual tourists: A Sahara desert experience. *Annals of Tourism Research* 70, 108–119.

Mujtaba, U. (2016) Ramadan: The month of fasting for Muslims, and tourism studies – mapping the unexplored connection. *Tourism Management Perspectives* 19 (B), 170–177.

Mullendore, N.D., Ulricch-Schad, J.D. and Prokopy, L.S. (2015) U.S. farmers' sense of place and its relation to conservation behavior. *Landscape and Urban Planning* 140, 67–75.

Munar, A.M. (2018) Foreword. In B.S.R. Grimwood, H. Mair, K. Caton and M. Muldoon (eds) *Tourism and Wellness: Travel for the Good of All?* (pp. xi–xiv). Lanham, MD: Lexington Books.

Murphy, J., Gretzel, U. and Hofacker, C. (2017) Service robots in hospitality and tourism: Investigating anthropomorphism. Paper presented at the *15th APacCHRIE Conference*, Denpasar-Bali, May.

Mzezewa, T. (2019) Saudi Arabia invites tourists: What you need to know. *The New York Times*, 27 September. See https://www.nytimes.com/2019/09/27/travel/saudi-tourist-visa-questions.html (accessed November 2019).

National Post (2015) Stop taking nude photos at ancient temples and historic sites: Cambodia to 'disrespectful' tourists. *National Post*, 27 February. See https://nationalpost.com/news/world/stop-taking-nude-photos-at-ancient-temples-and-historic-sites-cambodia-to-disrespectful-tourists (accessed November 2019).

Naugle, D.K. (2002) *Worldview: The History of a Concept*. Grand Rapids, MI: Eerdmans.

Nawijn, J. (2011) Determinants of daily happiness on vacation. *Journal of Travel Research* 50 (5), 559–566.

Nawijn, J. and Peeters, P.M. (2010) Travelling 'green': Is tourists' happiness at stake? *Current Issues in Tourism* 13 (4), 381–392.

Nawijn, J., Marchand, M.A., Veenhoven, R. and Vingerhoets, A.J. (2010) Vacationers happier, but most not happier after a holiday. *Applied Research in Quality of Life* 5 (1), 35–47.

Nawijn, J., Mitas, O., Lin, Y. and Kerstetter, D. (2012) How do we feel on vacation? A closer look at how emotions change over the course of a trip. *Journal of Travel Research* 52 (2), 265–274.

Neal, J.D. (2000) *The Effects of Different Aspects of Tourism Services on Travelers' Quality of Life: Model Validation, Refinement, and Extension*. Blacksburg, VA: Virginia Polytechnic Institute and State University.

Neri, C. (2006) Leadership in small groups: Syncretic sociality and the genius loci. *European Journal of Psychotherapy and Counselling* 8 (1), 33–46.

Newell-Hanson, A. (2018) 19 hotels that used to be churches, temples, and more. *Condé Nast Traveler*, 29 March. See https://www.cntraveler.com/galleries/2014-11-18/book-a-stay-in-a-converted-church-temple-and-more# (accessed November 2019).

Newman, W.J., Holt, B.W., Rabun, J.S., Phillips, G. and Scott, C.L. (2011) Child sex tourism: Extending the borders of sexual offender legislation. *International Journal of Law and Psychiatry* 34 (2), 116–121.

Nicolacopoulos, T. and Vassilacopoulos, G. (2004) On the other side of xenophobia: Philoxenia as the ground of refugee rights. *Australian Journal of Human Rights* 10 (1), 63–77.

Nielsen, K. (1993) Philosophy and Weltanschauung. *Journal of Value Inquiry* 27 (2), 179–186.

Niiiya, Y., Ellsworth, P. and Yamaguchi, S. (2006) Amae in Japan and the United States: An exploration of a 'culturally unique' emotion. *Emotion* 6 (12), 279–295.

Nikodemos and Makarios (1782) The Philokalia. Available online: https://holybooks.com/philokalia/

Nisbett, M. (2017) Empowering the empowered? Slum tourism and the depoliticization of poverty. *Geoforum* 85, 37–45.

Norberg-Schulz, C. (1979) *Genius Loci: Towards a Phenomenology of Architecture*. Los Angeles: Getty Publications.

Norman, A. and Pokorny, J.J. (2017) Meditation retreats: Spiritual tourism well-being interventions. *Tourism Management Perspectives* 24, 201–207.

Novelli, M., Morgan, N., Mitchell, G. and Ivanov, K. (2015) Travel philanthropy and sustainable development: The case of the Plymouth-Banjul challenge. *Journal of Sustainable Tourism* 24 (6), 824–845.

Nurse, K. (2004) Trinidad carnival: Festival tourism and cultural industry. *Event Management* 8 (4), 223–230.

Obrador, P. (2012) The place of the family in tourism research: Domesticity and thick sociality by the pool. *Annals of Tourism Research* 39 (1), 401–420.

Obrador-Pons, P. (2009) The Mediterranean pool: Cultivating hospitality in the coastal hotel. In P. Obrador-Pons, M. Crang and P. Travlou (eds) *Cultures of Mass Tourism: Doing the Mediterranean in the Age of Banal Mobilities* (pp. 91–110). Aldershot: Ashgate.

O'Connor, D. (2005) Towards a new interpretation of 'hospitality'. *International Journal of Contemporary Hospitality Management* 17 (3), 267–271.

Ogden, R.D., Hamilton, W.K. and Whitcher, C. (2010) Assisted suicide by oxygen deprivation with helium at a Swiss right-to-die organisation. *Journal of Medical Ethics* 36 (3), 174–179.

Ogletree, T.W. (2003) *Hospitality to the Stranger: Dimensions of Moral Understanding*. Louisville, KY: Westminster John Knox Press.

Oishi, S., Kesebir, S. and Diener, E. (2011) Income inequality and happiness. *Psychological Science* 22, 1095–1100.

O'Kane, R. (2018) 20 things airport employees are too polite to say out loud (but wish they could). *The Travel*, 26 July. See https://www.thetravel.com/20-things-airport-employees-are-too-polite-to-say-out-loud-but-wish-they-could/ (accessed August 2019).

Okumus, B., Okumus, F. and Kercher, B. (2007) Incorporating local and international cuisines in the marketing of tourism destinations: The cases of Hong Kong and Turkey. *Tourism Management* 28 (1), 253–261.

Olsen, D.H. (2003) Heritage, tourism, and the commodification of religion. *Tourism Recreation Research* 28 (3), 99–104.

Olsen, D.H. and Timothy, D.J. (2006) Tourism and religious journeys. In D. Timothy and D. Olsen (eds) *Tourism, Religion and Spiritual Journeys* (pp. 17–38). Abingdon: Routledge.

Omnia (2019) *Mountain Lodge: The Omnia*. See https://www.the-omnia.com/en/hotel/ (accessed August 2019).

Omondi, R.K. and Ryan, C. (2017) Sex tourism: Romantic safaris, prayers and witchcraft at the Kenyan coast. *Tourism Management* 58, 217–227.

Osborne, S. (2018) 'Tourist intrusion' – Ibiza draws up action plan as anger against holidaymakers grows. *Express*, 6 February. See https://www.express.co.uk/news/world/915271/ibiza-crackdown-british-holidaymakers-drunken-antisocial-behaviour-brits-abroad-tourism (accessed September 2019).

Osella, F., Stirrat, R. and Widger, T. (2015) Charity, philanthropy and development in Colombo, Sri Lanka. In B. Morvaridi (ed.) *New Philanthropy and Social Justice: Debating the Conceptual and Policy Discourse* (pp. 137–156). Bristol: Bristol University Press.

Pacek, A., Radcliff, B. and Brockway, M. (2019) Well-being and the democratic state: How the public sector promotes human happiness. *Social Indicators Research* 143 (3), 1147–1159.

Paget, S. and Regan, H. (2019) Ethiopia plants more than 350 million trees in 12 hours. CNN, 30 July. See https://edition.cnn.com/2019/07/29/africa/ethiopia-plants-350-million-trees-intl-hnk/index.html (accessed September 2019).

Palmer, G.E.H. and Ware, K.T. (eds) (2011) *The Philokalia, Vol. 4*. London: Faber & Faber.

Pan, S. (2011) The role of TV commercial visuals in forming memorable and impressive destination images. *Journal of Travel Research* 50 (2), 171–185.

Pan, Y., Weng, R., Xu, N. and Chan, K.C. (2018) The role of corporate philanthropy in family firm succession: A social outreach perspective. *Journal of Banking and Finance* 88, 423–441.

Papadopoulos, S. (2014) *O Archaggelos tou Souniou*. Limassol: Holy Church of Archaggelos Michael.

Paraskevaidis, P. and Andriotis, K. (2015) Values of souvenirs as commodities. *Tourism Management* 48, 1–10.

Pardwardhan, V., Ribeiro, M.A., Payini, V., Woosnam, K.M, Mallya, J. and Gopalakrishnan, P. (2019) Visitors' place attachment and destination loyalty: Examining the roles of emotional solidarity and perceived safety. *Journal of Travel Research* 59 (1), 3–21.

Parikiaki (2019) So bloody generous: Stranded passengers in Cyprus collect money for tearful flight crew who worked unpaid after travel company collapsed. *Parikiaki*, 30 September. See http://www.parikiaki.com/2019/09/so-bloody-generous-stranded-passengers-in-cyprus-collect-money-for-tearful-flight-crew-who-worked-unpaid-after-travel-company-collapsed/ (accessed October 2019).

Parkwell, C. (2019) Emoji as social semiotic resources for meaning-making in discourse: Mapping the functions of the toilet emoji in Cher's tweets about Donald Trump. *Discourse, Context and Media* 30, 100307.

Parrinello, G.L. (1993) Motivation and anticipation in post-industrial tourism. *Annals of Tourism Research* 20 (2), 233–249.

Paul, Apostle (n.d.) *A'. Epistle Paul towards Corinthians* (ιβ' 27–ιγ' 13).

Pearce, P.L. (2012) The experience of visiting home and familiar places. *Annals of Tourism Research* 39 (2), 1024–1047.

Pearce, P.L., Filep, S. and Ross, G.F. (2011) *Tourists, Tourism, and the Good Life*. New York: Taylor & Francis.

Pera, R., Viglia, C., Grazzini, L. and Dalli, D. (2019) When empathy prevents negative reviewing behavior. *Annals of Tourism Research* 75, 265–278.

Peters, M., Kallmuenzer, A. and Buhalis, D. (2019) Hospitality entrepreneurs managing quality of life and business growth. *Current Issues in Tourism* 22 (16), 2014–2033.

Petzet, M. (2008) *Genius loci – the spirit of monuments and sites*. In *16th ICOMOS General Assembly and International Symposium: 'Finding the Spirit of Place – Between the Tangible and the Intangible'*, Quebec, Canada.

Phillips, T. and Kaiser, A. (2019) Brazil must not become a 'gay tourism paradise', says Bolsonaro. *The Guardian*, 26 April. See https://www.theguardian.com/world/2019/apr/26/bolsonaro-accused-of-inciting-hatred-with-gay-paradise-comment (accessed August 2019).

Picard, D. and Robinson, M. (eds) (2006) *Festivals, Tourism and Social Change: Remaking Worlds*. Clevedon: Channel View Publications.

Pielichaty, H. (2015) Festival space: Gender, liminality and the carnivalesque. *International Journal of Event and Festival Management* 6 (3), 235–250.

Pietilä, M. and Fagerholm, N. (2016) Visitors' place-based evaluations of unacceptable tourism impacts in Oulanka National Park, Finland. *Tourism Geographies* 18 (3), 258–279.

Pizam, A. (2015) Is empathy essential for high-quality customer service? *International Journal of Hospitality Management* 49, 149–150.

Plog, S. (2001) Why destination areas rise and fall in popularity: An update of a Cornell Quarterly classic. *Cornell Hotel and Restaurant Administration Quarterly* 42 (3), 13–24.

Podoshen, J.S. (2013) Dark tourism motivations: Simulation, emotional contagion and topographic comparison. *Tourism Management* 35, 263–271.

Podoshen, J.S., Andrzejewski, S.A., Venkatesh, V. and Wallin, J. (2015) New approaches to dark tourism inquiry: A response to Isaac. *Tourism Management* 51, 331–334.

Poelmans, E. and Rousseau, S. (2016) How do chocolate lovers balance taste and ethical considerations? *British Food Journal* 118 (2), 343–361.

Pomfret, G. (2006) Mountaineering adventure tourists: A conceptual framework for research. *Tourism Management* 27 (1), 113–123.

Poria, Y., Beal, J. and Shani, A. (2019) Does size matter? An exploratory study of the public dining experience of obese people. *Journal of Hospitality and Tourism Management* 39, 49–56.

Porphyrios (2005) *Wounded by Love: The Life and Wisdom of Saint Porphyrios*. Evia: Denise Harvey.

Port Arthur Historic Site (2019) *Tickets and Tours*. See https://portarthur.org.au/tour/ghost-tour/ (accessed November 2019).

Porter, M.E. and Kramer, M.R. (2006) The link between competitive advantage and corporate social responsibility. *Harvard Business Review* 84 (12), 78–92.

Power, J. (2017) Ryanair crew told to sell at least eight scratch cards or face action. *The Irish Times*, 8 August. See https://www.irishtimes.com/business/transport-and-tourism/ryanair-crew-told-to-sell-at-least-eight-scratch-cards-or-face-action-1.3180262 (accessed August 2019).

Power, M.J. and Dalgleish, T. (2008) *Cognition and Emotion: From Order to Disorder* (2nd edn). Hove: Psychology Press.

Pratt, S. (2013) Minimising food miles: Issues and outcomes in an ecotourism venture in Fiji. *Journal of Sustainable Tourism* 21 (8), 1148–1165.

Pratt, S., Tolkach, D. and Kirillova, K. (2019) Tourism and death. *Annals of Tourism Research* 78, 102758.

Prayag, G., Mura, P., Hall, C.M. and Fontaine, J. (2016) Spirituality, drugs, and tourism: Tourists' and shamans' experiences of ayahuasca in Iquitos, Peru. *Tourism Recreation Research* 41 (3), 314–325.

Prayag, G., Hosany, S., Muskat, B. and Del Chiappa, G. (2017) Understanding the relationships between tourists' emotional experiences, perceived overall image, satisfaction, and intention to recommend. *Journal of Travel Research* 56 (1), 41–54.

Prentice, A.M. and Jebb, S.A. (1995) Obesity in Britain: Gluttony or sloth? *British Medical Journal* 311, 437–439.

Prihatna, A.A. and Abidin, H. (2005) *Muslim Philanthropy: Potential and Reality of Zakat in Indonesia: Survey Results in Ten Cities*. Jakarta: Piramedia.

Pritchard, A. and Morgan, N.J. (2001) Culture, identity and tourism representation: Marketing Cymru or Wales? *Tourism Management* 22 (2), 167–179.

Prose, F. (2003) *The Seven Deadly Sins: Gluttony*. Oxford: Oxford University Press.

Pruitt, D. and LaFont, S. (1995) For love and money: Romance tourism in Jamaica. *Annals of Tourism Research* 22 (2), 422–440.

Pullella, P. (2019) Rome bans sitting on Spanish Steps, puzzling hot, tired tourists. *Reuters*, 8 August. See https://www.reuters.com/article/us-italy-spanishsteps/rome-bans-sitting-on-spanish-steps-puzzling-hot-tired-tourists-idUSKCN1UY1YT (accessed September 2019).

Quoidbach, J., Taquet, M., Desseilles, M., de Montjoye, Y.A. and Gross, J.J. (2019) Happiness and social behavior. *Psychological Science* 30 (8), 1111–1122.

Radojevic, T., Stanisic, N., Stanic, N. and Davidson, R. (2018) The effects of traveling for business on customer satisfaction with hotel services. *Tourism Management* 67, 326–341.

Rahmani, K., Gnoth, J. and Mather, D. (2018) A psycholinguistic view of tourists' emotional experiences. *Journal of Travel Research* 58 (2), 192–206.

Raine, R. (2013) A dark tourist spectrum. *International Journal of Culture, Tourism and Hospitality Research* 7 (3), 242–256.

Ravenscroft, N. and Gilchrist, P. (2009) Spaces of transgression: Governance, discipline and reworking the carnivalesque. *Leisure Studies* 28 (1), 35–49.

Reis, H.T., Sheldon, K.M., Gable, S.L., Roscoe, J. and Ryan, R.M. (2000) Daily well-being: The role of autonomy, competence, and relatedness. *Personality and Social Psychology Bulletin* 26, 419–435.

Relph, E. (1976) *Place and Placelessness*. London: Pion.

Reynolds, T.E. (2010) Toward a wider hospitality: Rethinking love of neighbor in religions of the book. *Irish Theological Quarterly* 75 (2), 175–187.

Rich, M. (2018) How the Japanese fought suicide in the 'Sea of Trees'. *Independent*, 14 January. See https://www.independent.co.uk/news/long_reads/japan-suicide-logan-paul-youtube-vlogger-video-a8147516.html (accessed November 2019).

Riggio, R.E. and Reichard, R.J. (2008) The emotional and social intelligences of effective leadership: An emotional and social skill approach. *Journal of Managerial Psychology* 23 (2), 169–185.

Riley (2016) The Guardian. Available online: https://www.theguardian.com/global-development-professionals-network/2016/may/16/volunteers-stop-visiting-orphanages-start-preserving-families (accessed July 2020).

Ritschel, C. (2018) Tipping in America: How much should you really give when dining out in the US? *Independent*, 30 November. See https://www.independent.co.uk/life-style/tipping-america-how-much-us-percent-tip-waiter-restaurant-bar-a8657026.html (accessed August 2019).

Ritz-Carlton (2015) Dear Ritz-Carlton: WOW experiences for your employees? *Ritz Carlton Leadership Center*. See http://ritzcarltonleadershipcenter.com/tag/wow-story/ (accessed August 2019).

Ritzer, G. (2017) Hospitality and presumption. In C. Lashley (ed.) *The Routledge Handbook of Hospitality Studies* (pp. 233–246). Abingdon: Routledge.

Robb, E. (2009) Violence and recreation: Vacationing in the realm of dark tourism. *Anthropology and Humanism* 34 (1), 51–60.

Robbie, D. (2008) Touring Katrina: Authentic identities and disaster tourism in New Orleans. *Journal of Heritage Tourism* 3 (4), 257–266.

Robinson, R.N. (2019) Are things just too hot in the kitchen? Chefs' mental health and wellbeing. Presentation at *2019 International Conference: Navigating Shifting Sands: Research in Changing Times*, Melbourne, Australia.

Rodríguez-Campo, L., Braña-Rey, F., Alén-González, E. and Fraiz-Brea, J.A. (2019) The liminality in popular festivals: Identity, belonging and hedonism as values of tourist satisfaction. *Tourism Geographies* 22 (2), 229–249.

Rodriquez del Bosque, I. and San Martin, H. (2008) Tourist satisfaction: A cognitive-affective model. *Annals of Tourism Research* 35 (2), 551–573.

Rose, S. (2009) *The Soul After Death* (4th edn). Platina: St Herman of Alaska Brotherhood.

Rossiter, J. and Bellman, S. (2012) Emotional branding pays off. *Journal of Advertising Research* 52 (3), 291–296.

Rouby, C., Fournel, A. and Bensafi, M. (2016) The role of the senses in emotion. In H.L. Meiselman (ed.) *Emotion Measurement* (pp. 65–81). Cambridge: Woodhead.

Royal Caribbean (2016) Robot bartenders shake things up at sea. *Royal Caribbean*, 20 September. See https://www.royalcaribbean.com/blog/robot-bartenders-shake-things-up-at-sea/ (accessed August 2019).

Royal Caribbean (2019) *Symphony of the Seas – Things to Do*. See https://www.royalcaribbean.com/cruise-ships/symphony-of-the-seas/things-to-do (accessed September 2019).

Ruck, J. (2012) The destinations under threat from tourism – in pictures. *The Guardian*, 30 May. See https://www.theguardian.com/environment/gallery/2012/may/30/destinations-under-threat-tourism-in-pictures (accessed September 2019).

Rudgard, O. (2017) Visits to churches and cathedrals fall as fees and terror scare off tourists. *The Telegraph*, 18 August. See https://www.telegraph.co.uk/news/2017/08/18/visits-churches-cathedrals-fall-fees-terror-scare-tourists/ (accessed November 2019).

Ruiter, V. (2017) Hunter and gatherer hospitality in Africa. In C. Lashley (ed.) *The Routledge Handbook of Hospitality Studies* (pp. 247–259). Abingdon: Routledge.

Ryan, C. (2005) Ethics in tourism research: Objectivities and personal perspectives. In B.W. Ritchie, P. Burns and C. Palmer (eds) *Tourism Research Methods: Integrating Theory with Practice* (pp. 9–19). Wallingford: CABI.

Ryan, C. and Hall, C.M. (2001) *Sex Tourism: Marginal People and Liminalities*. London and New York: Routledge.

Ryan, C. and Trauer, B. (2005) *Sex and Tourism: Journeys of Romance, Love and Lust*. New York: Haworth Hospitality Press.

Ryff, C.D. (1989) Happiness is everything, or is it? Explorations on the meaning of psychological well-being. *Journal of Personality and Social Psychology* 57 (6), 1069–1081.

Ryff, C.D. and Singer, B.H. (2008) Know thyself and become what you are: A eudaimonic approach to psychological well-being. *Journal of Happiness* 9, 13–39.

Saarinen, J., Rogerson, C.M. and Hall, C.M. (2017) Geographies of tourism development and planning. *Tourism Geographies: An International Journal of Tourism Space, Place and Environment* 19 (3), 307–317.

Safyan, A. (2011) A call for international regulation of the thriving industry of death tourism. *Loyola of Los Angeles International and Comprehensive Law Review* 33 (287), 287–319.

Salisbury, J. (2018) Hellish holidays: Netflix's Dark Tourism reveals why people are visiting Japan's suicide forest and nuclear disaster zones. *The Sun*, 30 August. See https://www.thesun.co.uk/travel/7113989/netflixs-dark-tourism-reveals-why-people-are-visiting-japans-suicide-forest-and-nuclear-disaster-zones/ (accessed November 2019).

Salisbury, V. (2019) Easter Island statues are being defiled by nose-picking tourists. *New York Post*, 2 September. See https://nypost.com/2019/09/02/easter-island-statues-are-being-defiled-by-nose-picking-tourists/ (accessed November 2019).

Salzburgerland (2019a) *Alpine Summer: Fun and Relaxation in Nature*. See https://www.salzburgerland.com/en/alpine-summer/ (accessed August 2019).

Salzburgerland (2019b) *Farmhouse Holidays: Rustic Farms Nestled between Ski Slopes and Blooming Alpine Meadows*. See https://www.salzburgerland.com/en/farmhouse-holidays (accessed August 2019).

Sánchez, J.P. (2017) How the Parthenon lost its marbles. *National Geographic*, March/April. See https://www.nationalgeographic.com/history/magazine/2017/03-04/parthenon-sculptures-british-museum-controversy/ (accessed October 2019).

Sandals (2019) *Luxury Included Vacations for Two People in Love*. See https://www.sandals.com (accessed September 2019).

Sandstrom, G.M. and Dunn, E.W. (2014) Social interactions and well-being: The surprising power of weak ties. *Personality and Social Psychology Bulletin* 40, 910–922.

Saner, E. (2018) Employers are monitoring computers, toilet breaks – even emotions. Is your boss watching you? *The Guardian*, 14 May. See https://www.theguardian.com/world/2018/may/14/is-your-boss-secretly-or-not-so-secretly-watching-you (accessed August 2019).

Saveriades, A. (2000) Establishing the social tourism carrying capacity for the tourist resorts of the east coast of the Republic of Cyprus. *Tourism Management* 21 (2), 147–156.

Savoy-Sharm El Sheikh (2019) *Take your Pick of Convenience Savoy Resort Dining Packages*. See https://www.savoygroup-sharm.com/savoy/dining/packages (accessed September 2019).

Scammell, R. (2015) US tourists caught carving names into Rome's Colosseum. *The Guardian*, 8 March. See https://www.theguardian.com/world/2015/mar/08/us-tourists-caught-carving-names-into-colosseum-rome (accessed September 2019).

Schachter, S. and Singer, J.E. (1962) Cognitive, social and physiological determinants of emotional states. *Psychological Review* 69, 379–399.

Schlegelmilch, B.B. and Szőcs, I. (2017) Disaggregating corporate philanthropy: The impact of individual dimensions on customer-based corporate reputation. In C.L. Campbell (ed.) *The Customer is NOT Always Right? Marketing Orientations in a Dynamic Business World* (pp. 443–446). Cham: Springer.

Schupp, H., Stockburger, J., Codispoti, M., Junghofer, M., Weike, A. and Hamm, A. (2007) Selective visual attention to emotion. *Journal of Neuroscience* 27 (12), 1082–1089.

Seaton, A.V. (1996) Guided by the dark: From thanatopsis to thanatourism. *International Journal of Heritage Studies* 2 (4), 234–244.

Seaton, S. (2009) Purposeful Otherness: Approaches to the management of thanatourism. In R. Sharpley and P.R. Stone (eds) *The Darker Side of Travel: The Theory and Practice of Dark Tourism* (pp. 75–108). Bristol: Channel View Publications.

Sedgley, D., Pritchard, A., Morgan, N. and Hanna, P. (2017) Tourism and autism: Journeys of mixed emotions. *Annals of Tourism Research* 66, 14–25.

Sedikides, C. and Wildschut, T. (2016) Past forward: Nostalgia as a motivational force. *Trends in Cognitive Sciences* 20 (5), 319–321.

Sedikides, C., Wildschut, T., Routledge, C., Arndt, J., Hepper, E.G. and Zhou, X. (2015) To nostalgize: Mixing memory with affect and desire. *Advances in Experimental Social Psychology* 51, 189–273.

Semin, G.R. (2007) Grounding communication: Synchrony. In A.W. Kruglanski and E.T. Higgins (eds) *Social Psychology: Handbook of Basic Principles* (2nd edn) (pp. 630–649). New York: Guilford.

Semin, G.R. and de Groot, J.H.B. (2013) The chemical bases of human sociality. *Trends in Cognitive Sciences* 17, 427–429.

Shackley, M. (2001) *Managing Sacred Sites: Service Provision and Visitor Experience*. Boston, MA: Cengage Learning.

Shakeela, A. and Weaver, D. (2012) Resident reactions to a tourism incident: Mapping a Maldivian emoscape. *Annals of Tourism Research* 39 (3), 1337–1358.

Shakespeare, *2 Henry IV* 1.2.35–36.

Shamir, B. (1984) Between gratitude and gratuity: An analysis of tipping. *Annals of Tourism Research* 11 (1), 59–78.

Sharma, B. and Schultz, K. (2019) New Everest rules could significantly limit who gets to climb. *The New York Times*, 14 August. See https://www.nytimes.com/2019/08/14/world/asia/everest-climbing-rules.html (accessed August 2019).

Sharpley, R. (2005) Travels to the edge of darkness: Towards a typology of dark tourism. In C. Ryan, S. Page and M. Aitken (eds) *Taking Tourism to the Limits: Issues, Concepts and Managerial Perspectives* (pp. 217–228). Oxford: Elsevier.

Sharpley, R. and Jepson, D. (2011) Rural tourism: A spiritual experience? *Annals of Tourism Research* 38 (1), 52–71.

Sharpley, R. and Stone, P.R. (2009) (Re)presenting the macabre: Interpretation, kitschification and authenticity. In R. Sharpley and P.R. Stone (eds) *The Darker Side of Travel: The Theory and Practice of Dark Tourism* (pp. 109–128). Bristol: Channel View Publications.

Sharpley, R. and Sundaram, P. (2005) Tourism: A sacred journey? The case of ashram tourism, India. *International Journal of Tourism Research* 7 (3), 161–171.

Shaver, P., Morgan, H.J. and Wu, S. (2005) Is love a 'basic' emotion? *Personal Relationships* 3 (1), 81–96.

Shaw, A., Joseph, S. and Linley, P.A. (2005) Religion, spirituality, and posttraumatic growth: A systematic review. *Mental Health, Religion and Culture* 8 (1), 1–11.

Sheldon, P.J. and Park, S.Y. (2011) An exploratory study of corporate social responsibility in the U.S. travel industry. *Journal of Travel Research* 50 (4), 392–407.

Sheppard, V. (2010) Exploring the ethical standards of Alaska cruise ship tourists and the role they inadvertently play in the unsustainable practices of the cruise ship industry. In M. Lueck, P.T. Maher and E.J. Stewart (eds) *Cruise Tourism in Polar Regions* (pp. 101–118). London: Routledge.

Shields, R. (1990) The 'system of pleasure': Liminality and the carnivalesque at Brighton. *Theory, Culture & Society* 1 (1), 39–72.

Shim, C. and Santos, A.C. (2014) Tourism, place and placelessness in the phenomenological experience of shopping malls in Seoul. *Tourism Management* 45, 106–114.

Shondell Miller, D. and Gonzalez, C. (2013) When death is the destination: The business of death tourism – despite legal and social implications. *International Journal of Culture, Tourism and Hospitality Research* 7 (3), 293–306.

Shuo, Y.S.S., Ryan, C. and Liu, G.M. (2009) Taoism, temples and tourists: The case of Mazu pilgrimage tourism. *Tourism Management* 30 (4), 581–588.

Silk, J.B. (2006) Behaviour: Who are more helpful, humans or chimpanzees? *Science* 311, 1248–1249.

Simillidou, A. and Christou, P. (2018) A paradigm shift from 'emotional labour' to 'genuine emotional display' in the workplace. In B. Rowson and C. Lashley (eds) *Experiencing Hospitality*. Hauppauge, NY: Nova Science.

Simmonds, C., McGivney, A., Reilly, P., Maffly, B., Wilkinson, T., Canon, G., Wright, M. and Whaley, M. (2018) Crisis in our national parks: How tourists are loving nature to death. *The Guardian*, 20 November. See https://www.theguardian.com/environment/2018/nov/20/national-parks-america-overcrowding-crisis-tourism-visitation-solutions (accessed October 2019).

Singh, S. (2002) Love, anthropology and tourism. *Annals of Tourism Research* 29 (1), 261–264.

Sizer, S.R. (1999) The ethical challenges of managing pilgrimages to the Holy Land. *International Journal of Contemporary Hospitality Management* 11 (2–3), 85–90.

Skytrax World Airline Awards (2019) World's best airline cabin crew. *World Airline Awards*. See https://www.worldairlineawards.com/worlds-best-airline-cabin-crew-2019/ (accessed August 2019).

Slade, P. (2003) Gallipoli thanatourism: The meaning of ANZAC. *Annals of Tourism Research* 30 (4), 779–794.

Slattery, P. (2002) Finding the hospitality industry. *Journal of Hospitality, Leisure, Sport & Tourism* 1 (1), 19–28.

Small, J. (2016) Holiday bodies: Young women and their appearance. *Annals of Tourism Research* 58, 18–32.

Small, J. and Harris, C. (2012) Obesity and tourism: Rights and responsibilities. *Annals of Tourism Research* 39 (2), 686–707.

Smeekes, A. (2015) National nostalgia: A group-based emotion that benefits the in-group but hampers intergroup relations. *International Journal of Intercultural Relations* 49, 54–67.

Smith, D. (2016) Kenya burns largest ever ivory stockpile to highlight elephants' fate. *The Guardian*, 30 April. See https://www.theguardian.com/environment/2016/apr/30/kenya-to-burn-largest-ever-ivory-stockpile-to-highlight-elephants-fate (accessed August 2019).

Smith, H. (2018) Greece tourism at record high amid alarm over environmental cost. *The Guardian*, 3 June. See https://www.theguardian.com/world/2018/jun/03/greece-tourism-at-record-high-amid-alarm-over-environmental-cost (accessed October 2019).

Smith, M.K. and Diekmann, A. (2017) Tourism and wellbeing. *Annals of Tourism Research* 66, 1–13.

Smith, M. and Duffy, R. (2004) *The Ethics of Tourism Development*. New York: Routledge.

Smith, M. and Ram, Y. (2017) Tourism, landscapes and cultural ecosystem services: A new research tool. *Tourism Recreation Research* 42 (1), 113–119.

Smith, O. (2019) Revealed: The cities that could face an overtourism crisis in the next decade. *The Telegraph*, 12 June. See https://www.telegraph.co.uk/travel/city-breaks/cities-next-overtourism-battlegrounds/ (accessed August 2019).

Smith, S. (2015) A sense of place: Place, culture and tourism. *Tourism Recreation Research* 40 (2), 220–233.

Smith, V.L. (1992) Introduction: The quest in guest. *Annals of Tourism Research* 19 (1), 1–17.

Snyder, J. and Crooks, V.A. (2010) Medical tourism and bariatric surgery: More moral challenges. *American Journal of Bioethics* 10 (12), 28–30.

Sonmez, S., Apostolopoulos, Y., Theocharous, A. and Massengale, K. (2013) Bar crawls, foam parties, and clubbing networks: Mapping the risk environment of a Mediterranean nightlife resort. *Tourism Management Perspectives* 8, 49–59.

Soper, K., Ryle, M. and Thomas, L. (eds) (2009) *The Pleasures and Politics of Consuming Differently: Better than Shopping*. Palgrave Macmillan, London.

South China Morning Post (2017) China: 42 times tourists were caught behaving badly. *South China Morning Post*, 6 July. See https://www.scmp.com/news/china/article/2101412/39-times-tourists-were-caught-behaving-badly (accessed September 2019).

Speake, G. (2005) *Mount Athos: Renewal in Paradise*. New Haven, CT and London: Yale University Press.

Speake, G. and Ware, K. (2015) *Spiritual Guidance on Mount Athos*. Bern: International Academic.

Specht, S. and Kreiger, T. (2016) Nostalgia and perceptions of artwork, *Psychological Reports* 118 (1), 57–69.

Speed, C. (2008) Are backpackers ethical tourists? In K. Hannam and I. Atelievic (eds) *Backpacker Tourism: Concepts and Profiles* (pp. 54–81). Wallingford: CABI.

Spencer, A. and Bean, D. (2017) Female sex tourism in Jamaica: An assessment of perceptions. *Journal of Destination Marketing and Management* 6 (1), 13–21.

Squires, N. (2010) Vatican accused of hypocrisy over short skirts dress code. *The Telegraph*, 28 July. See https://www.telegraph.co.uk/news/worldnews/europe/vaticancityandholysee/7914540/Vatican-accused-of-hypocrisy-over-short-skirts-dress-code.html (accessed November 2019).

St Barnabas Orthodox Mission Kenya (2019) *St Barnabas Orthodox Orphanage and School*. See https://orthodoxmissionkenya.org/ (accessed August 2019).

St Basil (2016) *Saint Basic Collection: 4 Books*. Aeterna Press.

St Nektarios (2016) *Christology: Discovering Jesus Christ Through the Eyes of a Contemporary Saint*. Roscoe, NY: St Nektarios Monastery (Kindle edn).

St Nektarios of Aegina (2019) *Homilies by St Nektarios of Aegina, Vol. 1* (A. Skoubourdis and M. Agapi, trans). Athletis.

St Symeon Kolmogkorof (1998) *Elder Gabriel the Anchorite*. Attica: Botsis.

Stainton, H. (2016) A segmented volunteer tourism industry. *Annals of Tourism Research* 61, 256–258.

Sternberg, R.J. (ed.) (1990) *Wisdom: Its Nature, Origins, and Development*. Cambridge: Cambridge University Press.

Stevens, B. (2001) Hospitality ethics: Responses from human resource directors and students to seven ethical scenarios. *Journal of Business Ethics* 30 (3), 233–242.

Sthapit, E. and Coudounaris, D.N. (2018) Memorable tourism experiences: Antecedents and outcomes. *Scandinavian Journal of Hospitality and Tourism* 18 (1), 72–94.

Stoffelen, A. and Vanneste, D. (2015) An integrative geotourism approach: Bridging conflicts in tourism landscape research. *Tourism Geographies* 17 (4), 544–560.

Stone, A.A. and Mackie, C.E. (2013) *Subjective Well-being: Measuring Happiness, Suffering, and Other Dimensions of Experience*. Washington, DC: National Academies Press.

Stone, P. (2006) A dark tourism spectrum: Towards a typology of death and macabre related tourist sites, attractions and exhibitions. *Tourism* 52, 145–160.

Stone, P. and Sharpley, R. (2008) Consuming dark tourism: A thanatological perspective. *Annals of Tourism Research* 35 (2), 574–595.

Straits Times (2018) Bali to restrict tourist access in temples following incidents of disrespectful behaviour. *The Straits Times*, 27 September. See https://www.straitstimes.com/asia/se-asia/bali-to-restrict-tourist-access-in-temples-following-incidents-of-disrespectful (accessed November 2019).

Street, F. (2018) Can the world be saved from overtourism? *CNN Travel*, 3 October. See https://edition.cnn.com/travel/article/overtourism-solutions/index.html (accessed October 2019).

Strzelecka, M., Nisbett, G.S. and Woosnam, K.M. (2017) The hedonic nature of conservation volunteer travel. *Tourism Management* 63, 417–425.

Stunkard, A.J., LaFleur, W.R. and Wadden, T.A. (1998) Stigmatization of obesity in medieval times: Asia and Europe. *International Journal of Obesity* 22 (12), 1141–1144.

Su, L. and Swanson, S.R. (2017) The effect of destination social responsibility on tourist environmentally responsible behavior: Compared analysis of first-time and repeat tourists. *Tourism Management* 60, 308–321.

Sulek, M. (2010) On the modern meaning of philanthropy. *Nonprofit and Voluntary Sector Quarterly* 39 (2), 193–212.

Suleri, J. (2017) Experiencing hospitality and hospitableness. In C. Lasley (ed.) *The Routledge Handbook of Hospitality Studies* (pp. 326–336). Abingdon: Routledge.

Sullivan, P. (2019) Notre-Dame donation backlash raises debate: What's worthy of philanthropy? *The New York Times*, 26 April. See https://www.nytimes.com/2019/04/26/your-money/notre-dame-donation-backlash-philanthropy.html (accessed August 2019).

Sunder, K. (2017) 10 Reasons why I loved Kenya. *HuffPost Contributor Platform*, 23 November. See https://www.huffpost.com/entry/10-reasons-why-i-loved-ke_b_8621544 (accessed August 2019).

Susskind, J., Lee, D., Cusi, A., Feiman, R., Grabski, W. and Anderson, A. (2008) Expressing fear enhances sensory acquisition. *Nature Neuroscience* 11 (12), 843–850.

Swaine, M. (2019) *100% New Zealand: Food & Wine*. See https://www.newzealand.com/int/food-and-wine (accessed August 2019).

Swanson, K. (2015) Place brand love and marketing to place consumers as tourists. *Journal of Place Management and Development* 8 (2), 142–146.

Sykes, K. (1999) After the 'Raskol' feast: Youths' alienation in New Ireland, Papua New Guinea. *Critique of Anthropology* 19 (2), 157–174.

Tan, S.K, Tan, S.H., Kok, Y.S. and Choon, S.W. (2018) Sense of place and sustainability of intangible cultural heritage – the case of George Town and Melaka. *Tourism Management* 67, 376–387.

Tanaś, S. (2014) Tourism 'death space' and thanatourism in Poland. *Current Issues of Tourism Research* 3 (1), 22–27.

Tarlow, P.E. (2005) Dark tourism: The appealing 'dark' side of tourism and more. In M. Novelli (ed.) *Niche Tourism: Contemporary Issues, Trends and Cases* (pp. 47–57). Amsterdam: Elsevier.

Tasci, A.D.A. and Semrad, K.J. (2016) Developing a scale of hospitableness: A tale of two worlds. *International Journal of Hospitality Management* 53, 30–41.

Taushev, Archbishop A. (2014) *The Struggle for Virtue: Asceticism in a Modern Secular Society*. The Printshop of St Job of Pochaev. Jordanville, NY: Holy Trinity Monastery.

Telfer, E. (2013) The philosophy of hospitableness. In C. Lashley and A. Morrison (eds) *In Search of Hospitality: Theoretical Perspectives and Debates* (pp. 255–275). Oxford: Butterworth-Heinemann.

Teoh, M.W., Wang, Y. and Kwek, A. (2019) Coping with emotional labor in high stress hospitality work environments. *Journal of Hospitality Marketing and Management* 28 (8), 883–904.

Terkenli, T.S. (2002) Landscapes of tourism: Towards a global cultural economy of space? *Tourism Geographies* 4 (3), 227–254.

Terzidou, M., Scarles, C. and Saunders, M.N.K. (2018) The complexities of religious tourism motivations: Sacred places, vows and visions. *Annals of Tourism Research* 70, 54–65.

Thaddeus (2015) *Our Thoughts Determine our Lives: The Life and Teachings of Elder Thaddeus of Vitovnia* (A. Smiljanic, trans.). Platina, CA: St Herman of Alaska Brotherhood.

Thomas, M. (2015) Ryanair: Success before love. *Strategic Direction* 31 (8), 1–3.

Thompson, I. (2003) What use is the genius loci? In S. Menin (ed.) *Constructing Place: Mind and Matter* (pp. 66–76). London and New York: Routledge.

Tolkach, D., Pratt, S. and Zeng, C.Y. (2017) Ethics of Chinese and western tourists in Hong Kong. *Annals of Tourism Research* 63, 83–96.

Tomazos, K. and Cooper, W. (2012) Volunteer tourism: At the crossroads of commercialisation and service? *Current Issues in Tourism* 15 (5), 405–423.

Triantafillidou, A., Koritos, C., Chatzipanagiotou, K. and Vassilikopoulou, A. (2010) Pilgrimages: The 'promised land' for travel agents? *International Journal of Contemporary Hospitality Management* 22 (3), 382–398.

Tribe, J. (ed.) (2009) *Philosophical Issues in Tourism*. Clevedon: Channel View.

Tsaur, S.H. and Ku, P.S. (2017) The effect of tour leaders' emotional intelligence on tourists' consequences. *Journal of Travel Research* 58 (1), 63–76.

Tuan, Y.-F. (1974) *Topophilia: A Study of Environmental Perception, Attitudes, and Values*. Englewood Cliffs, NJ: Prentice Hall.

Tuan, Y.-F. (1979) Space and place: Humanistic perspective. In S. Gale and G. Olsson (eds) *Philosophy in Geography* (pp. 89–102). Dordrecht: D. Reidel.

Tuan, Y.-F. (1980) *Landscapes of Fear*. Oxford: Basil Blackwell.

Tuan, Y.-F. (1990) *Topophilia: A Study of Environmental Perceptions, Attitudes, and Values*. New York: Columbia University Press.

Tucker, H. (2016) Empathy and tourism: Limits and possibilities. *Annals of Tourism Research* 57, 31–43.

Tulloch, L. (2017) This is why Japan is the most polite nation on earth. *Traveller*, 15 April. See http://www.traveller.com.au/japan-the-most-polite-nation-on-earth-gv1oyp (accessed August 2019).

Tung, V.W.S. and Ritchie, J.B. (2011) Exploring the essence of memorable tourism experiences. *Annals of Tourism Research* 38 (4), 1367–1386.

Turkish Airlines (2019) *Our Social Responsibility Projects – 'Shall We Make a Snowman with You?'* See https://www.turkishairlines.com/en-jp/press-room/our-social-responsibility-projects/ (accessed 2019).

Tussyadiah, I.P. and Park, S. (2018) When guests trust hosts for their words: Host description and trust in sharing economy. *Tourism Management* 67 261–272.

Twenge, J.M., Sherman, R.A. and Lyubomirsky, S. (2016) More happiness for young people and less for mature adults: Time period differences in subjective well-being in the United States, 1972–2014. *Social Psychological and Personality Science* 7 (2), 131–141.

Umasuthan, H., Park, O.J. and Ryu, J.H. (2017) Influence of empathy on hotel guests' emotional service experience. *Journal of Services Marketing* 31 (6), 618–635.

Unger, R.H. and Scherer, P.E. (2010) Gluttony, sloth and the metabolic syndrome: A roadmap to lipotoxicity. *Trends in Endrocrinology & Metabolism* 21 (6), 345–352.

UN World Tourism Network on Child Protection (2019) *Ethics, Culture and Social Responsibility*. See https://www.e-unwto.org/doi/book/10.18111/9789284415588 and http://ethics.unwto.org/content/protection-children-tourism (accessed August 2019).

UNWTO (2019) *Global Code of Ethics for Tourism*. See https://www.unwto.org/global-code-of-ethics-for-tourism (accessed November 2019).

Upchurch, R.S. (1998) Ethics in the hospitality industry: An applied model. *International Journal of Contemporary Hospitality Management* 10 (6), 227–233.

Urry, J. (2016) The place of emotions within place. In J. Davidson, L. Bondi and M. Smith (eds) *Emotional Geographies* (pp. 77–83). London: Routledge.

Van Dijk, P., Smith, L.D.G. and Cooper, B.K. (2011) Are you for real? An evaluation of the relationship between emotional labour and visitor outcomes. *Tourism Management* 32 (1), 39–45.

Vantsos, M. and Kiroudi, M. (2007) An Orthodox view of philanthropy and church diaconia. *Christian Bioethics* 13 (3), 251–268.

Vargas-Sanchez, A., do Valle, P.O., da Costa Mendes, J. and Silva, J.A. (2015) Residents' attitude and level of destination development: An international comparison. *Tourism Management* 48, 199–210.

Varley, P.J. (2011) Sea kayakers at the margins: The liminoid character of contemporary adventures. *Leisure Studies* 30 (1), 85–98.

Verplanken, B. (2012) When bittersweet turns sour: Adverse effects of nostalgia on habitual worriers. *European Journal of Social Psychology* 42 (3), 285–289.

Veselka, L., Giammarco, E.A. and Vernon, P.A. (2014) The dark triad and the seven deadly sins. *Personality and Individual Differences* 67, 75–80.

Vignolles, A. and Pitchon, P. (2014) A taste of nostalgia: Links between nostalgia and food consumption. *Qualitative Market Research: An International Journal* 17 (3), 225–238.

Voigt, C., Brown, G. and Howat, G. (2011) Wellness tourists: In search of transformation. *Tourism Review* 66 (1–2), 16–30.

Von Schnurbein, G., Seele, P. and Lock, I. (2016) Exclusive corporate philanthropy: Rethinking the nexus of CSR and corporate philanthropy. *Social Responsibility Journal* 12 (2), 280–294.

Vreber, J.K., Zeigler-Hill, V., McCabe, G.A. and Baker, A.D. (2019) Pathological personality traits and immoral tendencies. *Personality and Individual Differences* 140, 82–89.

Walker, K. and Moscardo, G. (2016) Moving beyond sense of place to care of place: The role of Indigenous values and interpretation in promoting transformative change in tourists' place images and personal values. *Journal of Sustainable Tourism* 24 (8–9), 1243–1261.

Wallace, E. (2019) If you need a seat map to avoid crying babies, you're the problem. *CNN Travel*, 27 September. See https://edition.cnn.com/travel/article/parent-responds-japan-airlines-baby-map/index.html (accessed October 2019).

Walle, A.H. (1995) Business ethics and tourism: From micro to macro perspectives. *Tourism Management* 16 (4), 263–268.

Walsh, L.C., Boehm, J.K. and Lyubomirsky, S. (2018) Does happiness promote career success? Revisiting the evidence. *Journal of Career Assessment* 26 (2), 199–219.

Wang, C., Xu, H. and Li, G. (2017) The corporate philanthropy and legitimacy strategy of tourism firms: A community perspective. *Journal of Sustainable Tourism* 26 (7), 1124–1141.

Wang, K.L. and Groth, M. (2014) Buffering the negative effects of employee surface acting: The moderating role of employee-customer relationship strength and personalised services. *Journal of Applied Psychology* 2 (99), 341–350.

Wang, K., Lin, C.P., Chen, M.H. and Gillard, E. (2018) The impact of tourism firm's philanthropy decision on its business objective. *Tourism Economics* 24 (5), 503–509.

Wang, M., Chen, L., Su, P. and Morrison, A.M. (2019) The right brew? An analysis of the tourism experiences in rural Taiwan's coffee estates. *Tourism Management Perspectives* 30, 147–158.

Wang, Y.C., Qu, H. and Yang, J. (2019a) The formation of sub-brand love and corporate brand love in hotel brand portfolios. *International Journal of Hospitality Management* 77, 375–384.

Wang, Y.C., Ryan, B. and Yang, C.E. (2019b) Employee brand love and love behaviors: Perspectives of social exchange and rational choice. *International Journal of Hospitality Management* 77, 458–467.

Waxman, C.I. (2010) Beyond distancing: Jewish identity, identification, and America's young Jews. *Contemporary Jewry* 30 (2–3), 227–232.

Wearing, S. and McGehee, N.G. (2013) Volunteer tourism: A review. *Tourism Management* 38, 120–130.

Weaver, D.B. and Lawton, L.J (2007) 'Just because it's gone doesn't mean it isn't there anymore': Planning for attraction residuality. *Tourism Management* 28, 108–117.

Webber, E. (2019) Outraged officials in Bali warn tourists who disrespect sacred sites will be sent home or face 'purification rituals' after Czech man lifts up his girlfriend's skirt and splashes holy water on her bottom at Hindu temple. *Daily Mail*, 17 August. See https://www.dailymail.co.uk/news/article-7366461/Bali-officials-warn-tourists-disrespect-sacred-sites-sent-home.html (accessed November 2019).

Webster, J. (2003) An exploratory analysis of a self-assessed wisdom scale. *Journal of Adult Development* 10 (1), 13–22.

Weeden, C. (2002) Ethical tourism: An opportunity for competitive advantage? *Journal of Vacation Marketing* 8 (2), 141–153.

Weeden, C. (2015) Legitimization through corporate philanthropy: A cruise case study. *Tourism in Marine Environments* 10 (3–4), 201–210.

Weeden, C. and Boluk, K. (eds) (2014) *Managing Ethical Consumption in Tourism.* London: Routledge.

Weichselbaumer, D. (2012) Sex, romance and the carnivalesque between female tourists and Caribbean men. *Tourism Management* 33 (5), 1220–1229.

Weidenfeld, A. (2010) Iconicity and flagshipness of tourist attractions. *Annals of Tourism Research* 37 (3), 851–854.

Weiner, K. (2016) The trouble with tourists. *Harvard Political Review*, 30 November. See https://harvardpolitics.com/world/tourism-sexual-exploitation-of-children/ (accessed November 2019).

Weisskopf-Joelson, E. (1953) Some suggestions concerning Weltanschauung and psychotherapy. *Journal of Abnormal and Social Psychology* 48 (4), 601.

Wen, J., Meng, F., Ying, T., Qi, H. and Lockyer, T. (2018) Drug tourism motivation of Chinese outbound tourists: Scale development and validation. *Tourism Management* 64, 233–244.

Western Wall Heritage Foundation (2019) *The Western Wall: Reaching our Essence.* See https://english.thekotel.org/kotel/general_info/?itemid = %7BE7E0018D-6CC2-44DB-9CA3-AF340E5DCA72%7D (accessed November 2019).

White, M. (2019) Too many tourists: Should we limit visitor numbers to NZ? *NOTED*, 22 August. See https://www.noted.co.nz/money/money-economy/nz-tourists-should-we-limit-number-visitors (accessed October 2019).

Whitehead, J. (2018) Venice café sparks anger after charging tourists €43 for two coffees and two bottles of water. *Independent*, 7 August. See https://www.independent.co.uk/travel/news-and-advice/venice-st-marks-square-cafe-prices-tourists-san-marco-a8481376.html (accessed August 2019).

Wight, P. (1993) Ecotourism: Ethics or eco-sell? *Journal of Travel Research* 31 (3), 3–9.

Wilkins, H. (2011) Souvenirs: What and why we buy. *Journal of Travel Research* 50 (3), 239–247.

Wilkins, R. and Gareis, E. (2006) Emotion expression and the locution 'I love you': A cross-cultural study. *International Journal of Intercultural Relations* 30 (1), 51–75.

Willson, G.B. (2011) The search for inner peace: Considering the spiritual movement in tourism. *Journal of Tourism and Peace Research* 1, 16–26.

Wong, C.U.I., Ryan, C. and McIntosh, A. (2013) The Monasteries of Putuoshan, China: Sites of secular or religious tourism? *Journal of Travel & Tourism Marketing* 30 (6), 577–594.

Wong, J.Y. and Wang, C.H. (2009) Emotional labor of the tour leaders: An exploratory study. *Tourism Management* 30 (2), 249–259.

Wong, J., Newton, J.D. and Newton, F.J. (2014) Effects of power and individual-level cultural orientation on preferences for volunteer tourism. *Tourism Management* 42, 132–140.

Woodruff, P. (ed.) (2018) *The Ethics of Giving: Philosophers' Perspectives on Philanthropy*. Oxford: Oxford University Press.

Woodyatt, A. (2019) Amsterdam's mayor wants to reform red light district. *CNN Travel*, 4 July. See https://edition.cnn.com/travel/article/amsterdam-red-light-intl-scli/index.html (accessed August 2019).

Woosnam, K.M. (2012) Using emotional solidarity to explain residents' attitudes about tourism and tourism development. *Journal of Travel Research* 5 (3), 315–327.

Woosnam, K.M. and Norman, W.C. (2009) Measuring residents' emotional solidarity with tourists: Scale development of Durkheim's theoretical constructs. *Journal of Travel Research* 49 (3), 365–380.

Worth, H. (2006) Unconditional hospitality: HIV, ethics and the refugee 'problem'. *Bioethics* 20 (5), 223–232.

Wright, D.W.M. (2016) Hunting humans: A future for tourism in 2200. *Futures* 78, 34–46.

Wu, F. (2003) The (post-) socialist entrepreneurial city as a state project: Shanghai's reglobalisation in question. *Urban Studies* 40 (9), 1673–1698.

Yan, Q., Zhou, S. and Wu, S. (2018) The influence of tourists' emotions on the selection of electronic word of mouth platforms. *Tourism Management* 66, 348–363.

Yeh, S.-S., Chen, C. and Liu, Y. (2012) Nostalgic emotion, experiential value, destination image, and place attachment of cultural tourists. *Advances in Hospitality and Leisure* 8, 167–187.

Yeoman, I. and Mars, M. (2012) Robots, men and sex tourism. *Futures* 4, 365–371.

Yeung, S. (2004) Hospitality ethics curriculum: An industry perspective. *International Journal of Contemporary Hospitality Management* 16 (4), 253–262.

Ying, T. and Wen, J. (2019) Exploring the male Chinese tourists' motivation for commercial sex when travelling overseas: Scale construction and validation. *Tourism Management* 70, 479–490.

Yoon, Y. and Uysal, M. (2005) An examination of the effects of motivation and satisfaction on destination loyalty: A structural model. *Tourism Management* 26 (1), 45–56.

Young, M., Higham, J.E.S. and Reis, A.C. (2014) 'Up in the air': A conceptual critique of flying addiction. *Annals of Tourism Research* 49, 51–64.

Yu, J. (1998) Virtue: Confucius and Aristotle. *Philosophy East and West* 48 (2), 323–347.

Zarkia, C. (1996) Philoxenia: Receiving tourists – but not guests – on a Greek island. In J. Boissevain (ed.) *Coping with Tourists: European Reactions to Mass Tourism*. Oxford: Berghahn Books.

Zautra, A. (2003) *Emotions, Stress, and Health*. Oxford and New York: Oxford University Press.

Zhang, J.J., Wong, P.P.Y. and Lai, P.C. (2018) A geographic analysis of hosts' irritation levels towards mainland Chinese cross-border day-trippers. *Tourism Management* 68, 367–374.

Zhang, L., Chen, H. and Mo, S. (2016) Can corporate philanthropy be driven from the bottom to the top? *Academy of Management Proceedings* 1, 16576.

Zhang, X., Zhang, X. and Chen, X. (2017) Happiness in the air: How does a dirty sky affect mental health and subjective well-being?. *Journal of environmental economics and management* 85, 81–94.

Zheng, D., Ritchie, B.W., Benckendorff, P.J. and Bao, J. (2019) Emotional responses toward tourism performing arts development: A comparison of urban and rural residents in China. *Tourism Management* 70, 238–249.

Zhi, G.Y.J., Flaherty, G.T. and Hallahan, B. (2019) Final journeys: Exploring the realities of suicide tourism. *Journal of Travel Medicine* 26 (3).

Zoghbi-Manrique de-Lara, P., Aquiar-Quintana, T. and Suarez-Acosta, M.A. (2013) A justice framework for understanding how guests react to hotel employee (mis)treatment. *Tourism Management* 36, 143–152.

Zwolinski, J. (2019) Happiness around the world. In K.D. Keith (ed.) *Cross-Cultural Psychology: Contemporary Themes and Perspectives* (pp. 531–545). Malden, MA: Wiley-Blackwell.

Index

Aesthetics 60, 62
Agape 30, 34, 35, 36, 37, 38
Agape (actions) 35
Agathon: Glossary
Al-inlusive 101, 104, 105, 107
Alternative hedonism 63
Ambience 65
Anaesthetization 139
Anger 7, 19, 66, 67, 73, 74, 75
Animal sex tourism 32
Assisted suicide 140, 141
Aura 113, 137
Authenticity 90, 133, 134
Awe 66, 74

Benevolence 6, 21
Bestiality 32, 33
Body (human) 64, 67, 74, 83, 89, 103, 104, 123
Branding 52, 62, 65, 106

Carnival 94, 95, 96, 97
Carnivalesque 94, 95, 96, 97
Child sex tourism 32, 63
Comfort 16, 19, 22, 37, 40, 44, 70, 88, 104
Commodification 136, 137
Cosmos: Glossary; 2
COVID 24, 114, 134

Dark sites 132, 134, 139
Degradation (of environment/ landscape) 114, 121
Disgust 66, 67, 107
Doxey theory 23, 71
Dress codes 138

Egoism 43, 81
Emojis 74
Emotional exhaustion 92, 93

Emotional intelligence 70, 71
Emotional labour 18, 19, 70, 75, 92
Emotions 65, 66, 67, 68, 69, 70, 71, 72, 73, 74, 75, 76
Emotion (regulation) 68, 69
Empathy 34, 35, 70
Enclave 49, 96, 97
Entertainment 6, 35, 104, 105, 115, 139
Ethics 2, 3, 4, 7, 80, 124
Euthanasia 140, 141
Experiences (dark) 139
Experiences (individualistic) 88
Experiences (spiritual) 120, 125, 134
Experiences (thanatourism) 129, 130, 131, 132
Experiences (unexpected) 75

Fear 25, 66, 68, 75

Gastronomic (experiences) 6, 51, 73, 74, 100, 101, 103, 105
Genius loci 112, 113, 118, 119, 120, 121
Ghost tours 139
Gluttony 94, 99, 101, 102, 103, 106
Goodness 2, 4, 44, 46
Greed 2, 5, 35, 93, 102, 103

Hades 129
Happiness 66, 78, 79, 80, 81, 82, 83, 84, 85, 86
Happiness (subjective) 82
HASE (Holistic Agreeable Senses Experience) 61, 62
Hedonism 63, 89, 92, 94, 105, 106
Holocaust 6, 127, 131, 132, 134, 136
Hope 66, 76, 86
Hospitality 15, 16, 17, 18, 19, 20
Hospitality (commercial domain) 17
Hospitality (private domain) 16
Hospitality (sociocultural domain) 16

Infantilisation (tourist) 104

Joy 8, 74, 75, 79, 81, 86, 88

Kalon: Glossary; 38
Kindness 2, 6, 15, 22, 23, 26, 29, 34, 37, 39, 41, 84, 138

Life satisfaction 78, 80, 82, 85
Liminality 95, 96, 97
Love 30, 31
Love (forms) 33, 34
Luxury 51, 61, 93, 100, 104

Major Depressive Disorder 86, 87, 92
Memorabilia 62, 65, 121
Mistreatment (of employees) 55
Moderation 99
Morals 3, 4, 124
Morphology 75
Motivation (tourist) 60, 69, 100, 126
Motivation (pull factors) 100
Motivation (push factors) 100
Mount Athos 28, 88, 125, 126, 128

Neoteny 123
Nostalgia 64, 65

Obesity 102, 105
Ometenashi: Glossary; 17
Overconsumption 99, 101, 102, 103, 104, 105, 106, 107
Overindulgence 94, 102, 103, 104, 106
Overtourism 23, 24, 100, 121
Overtourism (unethical tourist behavior) 134

Paedomorphosis 123
Paramorphosis 115
Philagathy 38
Philanthropy 44, 53
Philanthropy (critical) 52
Philanthropy (hypocritical) 52
Philanthropy (organisational) 48
Philanthropy (private) 45
Philanthropy (ultimate) 52
Philosophy: Glossary; 1; 2, 4, 5
Philoxenia 20, 21, 22, 23, 24, 25, 26, 27, 28, 29
Philoxenia (employee influences) 27

Philoxenia (exogenous influences) 23
Philoxenia (organisational influences) 25
Philoxenia (tourist/guest influences) 26
Place (affection) 111
Place (alteration) 113
Place (capsules) 120
Place (genius loci) 113
Place (placelessness) 113
Place (of atrocity) 129, 133, 135, 136
Place (sacred) 134, 136, 137, 138
Pride 75
Profiteering (conduct/attitude) 28, 41
Psychology (and well-being) 77

Quality standards 19

Relaxation 85, 90, 92, 100

Sacred sites (categories) 127
Sadness 66, 73, 75, 132
Satisfaction 37, 45, 59, 61, 67
Saturnalia 95
Selfies 97, 134, 135, 136
Sense of place 112
Senses (human) 59, 60, 61, 62, 63, 64, 65
Senses (HASE) 61, 62
Setting (utopic) 73
Sex tourism 31, 32
Shame 66, 103
Silence 2, 87, 89, 90
Slum tourism/slum tours 50, 51, 53
Smile 16, 18, 27, 35, 37, 67, 81
Social media 7, 32, 37, 70, 74, 87, 98, 136, 138
Soul 64, 79, 89, 113
Space (liminal) 95
Spirit (of places) 113
Surprise 23, 45, 66, 88

TALC (Tourism Area Life Cycle) 115
Tangibility 62
Thanatourism 129, 130, 131, 132
Topophilia 66, 111
Tourism (genocide) 129, 131
Tourism (religious) 124
Tourism (spiritual) 124
Tourism experiences (hedonistic) 103, 104, 105
Tourist (emotions) 65
Tourist (senses) 59

Tourist behavior (unethical) 26, 97
Tourist behavior (unruly) 27, 62, 94
Tourist philanthropy 46

Virtual experiences 59
Virtues: Glossary; 4, 5, 103, 138
Voluntourism (volunteer tourism) 35, 46, 47, 54

Well-being 77, 78, 79, 80, 81

Well-being (psychological) 78
Well-being (subjective) 78
Wellness 77, 87, 124, 127
Weltanschauung: Glossary; 2
Wisdom 1, 2, 4

Xenos 20

Zoophilia 32
Zoophiles 33

Lightning Source UK Ltd.
Milton Keynes UK
UKHW022030170121
377023UK00011B/284